AMERICAN
WARFIGHTER

AMERICAN WARFIGHTER

BROTHERHOOD, SURVIVAL, AND UNCOMMON VALOR IN IRAQ

2003-2011

J. PEPPER BRYARS

American Warfighter:
Brotherhood, Survival, and Uncommon Valor in Iraq, 2003-2011

Copyright 2016 by J. Pepper Bryars

www.jpepperbryars.com

Hardcover ISBN: 978-0-9913248-4-2
Paperback ISBN: 978-0-9913248-5-9
Ebook ISBN: 978-0-9913248-6-6
Audiobook ISBN: 978-0-9913248-7-3

Cover Photograph: Soldiers from Battery B, 3rd Battalion, 320th Field Artillery Regiment, 101st Airborne Division, pause at the end of a patrol near Wynot, Iraq, September 2005. Courtesy of the U.S. Army.

Book design by Maureen Cutajar
Cover design by Jesus Cordero

DEDICATION

In memory of the first and last Americans to die in the Iraq War: Therrel Shane Childers, killed in action on March 21, 2003, in southern Iraq, and David Emanuel Hickman, killed in action on November 14, 2011, in Baghdad.

"War is brutish, inglorious, and a terrible waste. The only redeeming factors were my comrades' incredible bravery and their devotion to each other."

— E. B. Sledge,
With the Old Breed at Peleliu and Okinawa

"We have good corporals and good sergeants and some good lieutenants and captains, and those are far more important than good generals."

— William Tecumseh Sherman,
wartime correspondence, *1864*

"For all the brilliance of the generals, in the end success or failure in Operation Overlord came down to a relatively small number of junior officers, noncoms, and privates or seamen."

— Stephen Ambrose,
D-Day: June 6, 1944: The Climactic Battle of World War II

CONTENTS

AUTHOR'S NOTE

The book you're about to read is true. Through dozens of hours of interviews and two years of research, I worked closely with ten highly decorated warfighters to weave their individual experiences into a narrative spanning the entire Iraq War. Many outside sources are referenced, but the stories remain faithful to each warfighter's unique perspective. They speak for themselves, and with a remarkable degree of openness.

No pseudonyms are used, and the acts of valor of each warfighter were independently confirmed with official records. The military conducts thorough investigations before approving awards of the caliber mentioned in this book. Simply put, these heroes are the real deal.

Some significant information is shared, but none of it is classified. Officials at the Pentagon cleared the text, and their request to remove sensitive information was honored.

The labels for those we fought are used interchangeably. Unless the names are mentioned for a specific purpose, they're simply called "enemy fighters" or "insurgents" to ease the reader's burden of differentiating between the many groups we fought in Iraq. The ranks and units mentioned are accurate for the periods in which the events took place, and any mistakes are my own.

These warfighters have entrusted us with a precious gift—their memories. May we never forget.

—*J.P.B.*

PART I

THE INVASION, 2003

CHAPTER ONE

A TIME OF WAR

In early 2003, a city of warriors had arisen from the sands of the Kuwaiti desert. In only a few short weeks, where there had once been nothing but a flat, dry, and barren wasteland, there now stood row upon row of tents, vehicles, weapons, ammunition, and supplies of every sort. The silence and stillness were replaced with the cracking and thunderous sounds of firing ranges, the hum of diesel generators and supply convoys, and the cacophony of tens of thousands of American warfighters who were readying themselves to cross an unseen yet deadly line in the sand.

One of those Americans was Second Lieutenant Therrel Shane Childers, a platoon leader in Alpha Company, 1st Battalion, 5th Marine Regiment, or 1/5 Marines, out of Camp Pendleton, California.

Childers was the thirty-year-old son of a career military man whose accent and values were a blend of the many places he had lived as a boy: West Virginia, Mississippi, Wyoming, the Carolinas, California, and Puerto Rico. He was a sergeant before earning his officer's commission at the Citadel, the prestigious military academy in South Carolina. Childers was the quintessential Marine, and all of his life, combined with years of professional military training and education, had been leading to this moment. Childers was a warrior who was finally at war.

One of the Marines in his company, Sergeant Bradley Nerad, would later explain Childers' dedication to the profession of arms in *Men's Journal.*

3

"He was living for that day," Nerad said. "Man ... he was in it for this."

But Nerad also described a strange moment that happened shortly before the invasion kicked off. He and Childers were in the middle of one of the military's perpetual "hurry up and wait" moments, passing the time by talking about what the war would be like. At one point Childers stared into the desert and said, in an unfamiliar, foreboding tone, "I probably won't come back."

Nerad responded, "Hey, c'mon."

"I don't know, man," Childers said. "I just don't think I'm coming home."

On the other side of the world, America was ready for the war to begin. A day before the invasion *ABC News* reported that spot polling showed that "71 percent say they'd support war, up a dozen points from a week ago" and that there was "broad consensus that diplomacy has run its course." Perhaps speaking for most Americans, musician Toby Keith was quoted on the cover of *Country Weekly* as saying, "If my son is going to war, I want the military kicking ass."

That's exactly what Childers and his Marines wanted to do. Shortly before the invasion, Childers wrote home telling his parents, "We're ready. Every day we just get harder and tighter, more disciplined."

Then finally, after months of being on a tightly held leash, the Marines of Alpha Company were set loose to storm across the desert on March 21, 2003. Their mission: to secure a massive pumping station in the vast Rumaila oil fields of southern Iraq. Our military planners believed the Iraqi army would sabotage the facility in an effort to slow the coalition's advance.

The article in *Men's Journal* described the scene: "Childers roars across the Rumaila oil fields in a large convoy—some 200 men riding in several dozen tracks, Humvees, and tanks. His lip bulges with Copenhagen as he alternately fidgets with a map and squints at the landscape. The country, what little Childers can see of it in the darkness, is flat and featureless and marked only by the occasional piece of derelict petroleum equipment half-swallowed by the sand. Up ahead trench oil fires burn in large arcs around the pumping station complex, laying down a mantle of noxious smoke that has rendered his night-vision equipment useless."

The convoy continued until it reached the pumping station's industrial complex. Shortly after dawn the Marines successfully secured the under-defended facility and took several Iraqi prisoners. It was a swift victory, and Childers called his men together to discuss the morning's events and give orders for their next actions.

The Iraq War would eventually last 3,194 days and cost an average of 1.4 American lives per day, but at that moment, the conflict was less than forty-eight hours old and was about to claim its first American life.

As Childers assembled his men, a Toyota truck full of Iraqis fleeing the facility barreled down a road. As they passed, the driver fired a wild burst from his rifle, striking Childers inches below his body armor. He fell to his knees, and then rolled over, curling himself on the ground.

"I'm hit," he moaned. "In the gut."

Marines rushed to his aid, performing CPR while the Navy corpsman tried to keep him alive long enough for a helicopter to evacuate their lieutenant to a combat hospital in Kuwait. The wound was too severe, however, and the internal bleeding slowly sapped Childers of life. His men labored valiantly to save their leader, but nothing worked. Around nine o'clock that morning, Second Lieutenant Therrel Shane Childers died, and became the first American killed in action during the Iraq War.

The road that led Childers and the men of Alpha Company, 1/5 Marines, to that lonely pumping station began in earnest more than a year earlier, in January of 2002 when senior Bush Administration officials began to seriously contemplate war with Iraq.

"Everybody was convinced that there'd be a follow-on to 9/11," Vice President Dick Cheney said years later in a BBC documentary about the war. "The place where we thought the biggest threat [lay] was Saddam Hussein and Iraq."

Cheney wasn't alone in his assessment. Intelligence officials across the Western world became convinced that Iraqi president Saddam Hussein was growing an immense arsenal of chemical, biological, and possibly even nuclear weapons. It was a clear threat to a vital region of the world, many believed, and possibly a direct threat to the United States. Hussein had repeatedly denied international weapons inspectors access to his military and government facilities, and had increased his threatening rhetoric during a time when other dictators around the Middle East were remaining silent in the wake of the September 11th attacks.

Adding to this perceived threat was Iraq's history of using chemical weapons. Exactly fifteen years before the invasion, in March of 1988, the regime henchman who would eventually become known as "Chemical Ali," Ali Hassan al-Majid, ordered a gas attack on the small town of Halabja in northern Iraq.

"When you hear people shouting the words 'gas' ... terror begins to take hold, especially among the children and the women," one survivor told Radio Free Iraq. "Your loved ones, your friends, you see them ... falling like leaves to the ground ... birds began falling from their nests; then other animals, then humans ... whoever had too many children to carry on their shoulders, they stayed in the town and succumbed to the gas."

More than five thousand men, women, and children are believed to have died in that 1988 attack. Photographs eventually made their way out of the town, showing the dead as they fell. In one picture, a man lay dead in the street, a scarf tied around his face in a failed defense against the gas. In his arms he clutched his dead son, no more than a few months old, wrapped tightly in swaddling clothes. The infant's mouth still gaped open, frozen in the moment it gasped through the burning chemicals for its final breath of life.

Our national security officials had not forgotten that attack, and many were deeply worried about what could happen if terrorists obtained those chemicals from Hussein's alleged stockpile. Our national defenses had been taken by surprise in New York City, and many feared it could happen again. The notion of a preemptive strike against known threats became not only an acceptable justification for war, but also quite nearly an automatic demand for military action considering 9/11. After months of continued stalling from Hussein's regime, the United States moved closer to all-out conflict.

"We know that Saddam Hussein pursued weapons of mass murder even when inspectors were in his country. Are we to assume that he stopped when they left?" asked President George W. Bush during his address to the United Nations General Assembly in September 2002. "The history, the logic, and the facts lead to one conclusion: Saddam Hussein's regime is a grave and gathering danger."

In early October 2002 a joint resolution authorizing the invasion of Iraq was introduced in Congress. The debate sharpened and support for action solidified. After months of absorbing testimony and listening to world leaders debate the issue, the American public had decided that the status quo wasn't acceptable. Most believed that Hussein's continued control of Iraq posed a clear threat to the United States and that it was worth risking military action to remove him from power. At the time, many believed it would have been irresponsible, and clearly dangerous, not to disarm Hussein.

"The time for denying, deceiving, and delaying has come to an end," President Bush said from Ohio in early October 2002. "Saddam Hussein must disarm himself—or, for the sake of peace, we will lead a coalition to disarm him."

On the afternoon of Thursday, October 10, 2002, the resolution authorizing the war passed the House by a vote of 296–133. Only six Republicans voted against the measure, and 40 percent of Democrats voted in favor. It passed the Senate shortly after midnight the following morning by a vote of 77–23. President Bush had the overwhelming approval of the people's representatives to invade Iraq and forcibly disarm the Hussein regime.

Over the next several months, American troops and others from what became known as a "coalition of the willing" began deploying to staging bases in Kuwait to prepare for the invasion. More warnings and diplomatic pleas were made to the Hussein regime to disarm and open Iraq to full and unfettered outside inspection. The warnings were ignored, and our nation was resolved for war.

On the evening of March 17, 2003, President Bush gave a televised address from the White House in which he delivered an ultimatum for the world to hear.

"All the decades of deceit and cruelty have now reached an end," Bush said. "Saddam Hussein and his sons must leave Iraq within forty-eight hours. Their refusal to do so will result in military conflict commenced at a time of our choosing."

America was an unstoppable force, and Saddam Hussein believed that his regime was an immovable object. There was nothing else to be done. As the book of Ecclesiastes says, "There is an appointed time for everything, and a time for every affair under the heavens."

The spring of 2003 became "a time of war" in Mesopotamia. And so war came ... and the invasion began.

CHAPTER TWO

"THIS IS WHAT CUSTER MUST HAVE FELT LIKE."

JUSTIN LeHEW
NAVY CROSS, *NASIRIYAH*

*A*fter months of debate and preparation, on the evening of Wednes-day, March 19, 2003, President Bush delivered a televised speech from his desk in the Oval Office announcing that the war had begun: "On my orders, coalition forces have begun striking selected targets of military importance to undermine Saddam Hussein's ability to wage war. These are opening stages of what will be a broad and concerted campaign."

After briefly noting that thirty-five countries were part of the coalition effort, the president turned his words to our warfighters: "The peace of a troubled world and the hopes of an oppressed people now depend on you. That trust is well placed. The enemies you confront will come to know your skill and bravery. The people you liberate will witness the honorable and decent spirit of the American military."

The size of the initial invasion force depended upon which troops were counted—those who actually crossed into Iraq, around 150,000, or those who supported the invasion aboard ships or air bases, which brought the total to nearly 500,000. When they were counted mattered, too, since preparation for the invasion lasted several months. But whatever the actual size of the force, the number of American warfighters on the ground inside of Iraq would be far fewer than the "several hundred thousand" that the Army's chief of staff said would be required to occupy the country once Saddam was defeated.

It took our commanders years to realize that more troops were needed, partly because many of their training scenarios showed that Americans weren't supposed to be fighting this kind of old-fashioned war, where troops hunted the enemy house to house and fought like their fathers had done in Vietnam, and how their grandfathers had done in Korea and in World War II. The previous decade saw an explosion in technology that the Pentagon labeled a "Revolution in Military Affairs" and signaled a change in the way America would fight its wars. It was supposed to be a lightning-fast, high-tech war in which having an overwhelming number of troops shouldn't have mattered. More smart bombs from the air and less boots on the ground, the generals said. Precision over brute force. Networks, not battlefields. Technological innovations certainly played a supporting role in the Iraq War, but the old kind of warrior, fighting the old kind of way, and with the old kind of tools—courage, and a rifle—would be needed to win this supposedly new kind of war.

One of the American warfighters who was about to invade Iraq on that March evening was thirty-three-year-old Gunnery Sergeant Justin LeHew, a native of Columbus Grove, Ohio. He was cut from the same cloth that produced the "Old Breed" of Marines who won World War II. Before ever becoming a Marine, however, LeHew had first tried to enlist in the U.S. Air Force straight out of high school but was told at the Military Entrance Processing Station that he couldn't join that branch because he was colorblind. It was a blow; he felt completely dejected. After wandering around the enlistment station for a little while, LeHew, then only seventeen years old, sat on a bench and waited for the bus that would take the enlistees back home. He was looking down, mulling over the disappointing news and wondering what he was going to do with his life, when a pair of sharply creased blue pants with a red stripe down the side came into his field of vision. "What's your problem? You look like you just lost your favorite dog," asked a Marine gunnery sergeant named Spehar. LeHew began to explain when the sergeant interrupted. "That's not the way it works here," Spehar said. "You get on your feet when you talk to me."

At that very moment, everything changed for LeHew. He knew he was going to be a United States Marine.

Years later, at the outset of the Iraq War, LeHew was a platoon sergeant in charge of several assault amphibious vehicles, commonly known as "AmTracs"

*because of their original name: amphibious tractors. "I lucked into my job,"
he said, describing his time at boot camp in 1988. "The night before boot
camp graduation at Parris Island, the senior drill instructor sat the whole
platoon down, just like in 'Full Metal Jacket,' and announced our names
and we'd stand up. He said, 'LeHew.' I jumped up and said, 'Sir, yes, sir!'
He said, '1833: AmTrac crewman. Congratulations.' I remember sitting
down and was like, 'Gosh, what is that?' I had no idea what I wanted to
be, but when he said AmTrac crewman, I remember thinking, 'Shit, I'm
going to be at some train depot loading Amtrak trains with supplies to
send to combat troops. You've got to be kidding me.' Then they showed me:
an AmTrac is a 26-ton floating tank that has a 40 millimeter grenade
launcher, a .50-caliber machine gun, and can carry 21 combat-loaded
troops from ship to shore and then to inland objectives. I was hooked."*

JUSTIN LEHEW: I was with Alpha Company, First Battalion, Second
Marines, and the operational code for our overall organization was
Task Force Tarawa. During the invasion, Lieutenant Keith Brenize
and I were responsible for twelve AmTracs, forty Marines, and one
Navy corpsman.

We only had a few short months to train them because we knew
something was coming. So I took our vehicles to the field and I made
those Marines live in the dirt at Camp Lejeune, North Carolina, to learn
their trade. We put them through tasks, evaluated them, and then moved
crews around. Then we learned unit cohesion. That's part of the reason
that we were successful in the most hostile and extreme circumstances.

War is ugly. I was in Desert Storm and ran POWs as a young
Marine. For the most part, the Iraqis in that war simply threw up
their arms and were like, "Hey, man, we don't want this." Hundreds
of thousands of them. Stacks and stacks of them in our vehicles.
There were so many surrendering that you would point in the
southerly direction to say, "Just keep walking that way," because you
couldn't fit them all in your vehicle.

But I remember the looks on their faces, and I could tell the officers
from the enlisted. I could see the way their attitude was about war. I

could sense, even as a young Marine, that the fight wasn't finished. They had just come off of fighting the Iran-Iraq war, from 1980 to 1988. And now Desert Storm was only three years after that. They've been living in a warfare lifestyle for years thinking, "My brother's going to kill me tomorrow." I could see that they were angry and they were tired. They were discouraged because they saw the full troop strength and combat power of the United States Armed Forces. The B-52 bombers, the constant bombardment, and the coordinated effort in the attack ... they looked around and were simply demoralized.

It was more than that, though. They grew up in war, and for thousands of years that region had fostered a warring culture. Borders have changed over the years, but that mentality didn't leave. So when I looked at them, I could tell that they didn't have a lot of fight left, but I could also see that this wasn't finished.

I believe the reason they didn't have a lot of fight left in them, quite frankly, was that it's pretty easy to invade somebody else's country, Kuwait in this instance, and then give that ground back and retreat into your own country and say, "Hey, I'm sorry I did that." We didn't pursue them up into Baghdad in 1991. All they had to do was get north of that border, and they were good.

The general consensus of the leadership in 2003 was that the same shit was going to happen this time, too. They thought that as soon as we showed up with all of our combat power, as soon as we flew B-52s and Stealth bombers and started pounding the Iraqis, the same thing was going to happen. They thought that the Iraqis were going to throw their hands up and they were not going to fight.

I knew there was no way that was going to happen inside Iraq. We weren't pushing the Iraqis out of somebody else's country; we were invading their country now. Think about it; there's nobody that's going to come into the United States and roll an army through Kansas. Those farmers aren't going to throw their hands up and give up their cattle, livestock and everything without a fight. That's the difference. The difference in 1991 was, "Get them out of somebody else's country and put them back in theirs." In 2003, it was, "We're coming after you in your own country." I knew there weren't going to be people throwing up their hands. They were going to fight. So we were going to train and do whatever it took to prepare for that fight.

My job was to train my unit well enough and bring as many of them home as possible. So I trained that mentality into those young Marines, to identify scenarios, to know how to handle the indigenous population, to make sure that they weren't burning the villages because, in the minds of the young Marines, these were the people who flew those planes into the World Trade Center, and they were pissed.

We wanted retaliation. That was the mentality of a lot of people going into Iraq: "We're coming to get you because you did this." I can't remember talking about weapons of mass destruction and getting all wrapped up in any of that stuff. When we got over there, seven thousand miles away from home, nobody gave a rat's ass who the president was, nobody cared anything about politics, and who's Republican and who's Democrat. All they cared about was surviving, accomplishing the mission, and getting the hell out of there.

We couldn't have prepared any better than we did, but it's like a team an hour before any football game—you go with what you've got. So from Camp Lejeune, North Carolina, we floated over to the Middle East on the USS *Portland*, which was an Anchorage-class dock landing ship.

On the way over we couldn't use our AmTracs, of course, so we found other ways to train, like using Matchbox cars to teach tactics. Our unit never laid in their racks on the ship trying to figure out what they were going to do. Their days were highly structured. They were up in the morning and doing everything they were doing back in Camp Lejeune, but they were doing it on a Navy ship in the middle of the ocean, going to war. They had physical training, and would work until 6 or 7 p.m. every night, religiously. We moved ammunition from the front of the ship to the back. We did gun drills and whatever else was needed to make them into a cohesive unit, which is one of the reasons we didn't have our butts handed to us at the Battle of Nasiriyah.

We offloaded and started moving our vehicles to our staging

points in Kuwait. I was very confident that we were taking the best that America had to offer—the best Marines, the best soldiers, sailors, and airmen, that we could put on that battlefield. I was very confident in the team that was going into the fight.

We started staging on March 19th, started moving into Iraq on the 21st, and when we finally hit the triggers it was the 23rd. We lifted a company of infantry in our AmTracs, roughly 140 to 160 men with all their weapons and gear: everything from their personal 9-millimeter sidearms to M16s, M240 machine-guns and M249 SAWs [squad automatic weapon]. Then we had the mortar men and carried their 60-millimeter and 81-millimeter mortars. At that time they had these assault men with thermobaric rocketry to blow up tanks. So we were carrying all of that inside our vehicles, with piles and piles of ammunition. It was cramped for the next couple hundred miles.

Justin LeHew atop his AmTrac shortly before the invasion,
March 2003 (*J. LeHew*)

I don't think that anyone was prepared for how rapidly we moved into and through Iraq. We covered hundreds of miles so fast. The only time we stopped was when a refueler pulled up alongside and fueled the vehicles. It was normally just a big gas truck, and our tanks ate more gas than anybody so we had to constantly refuel them. Our AmTracs can travel three hundred miles on land completely fueled, but all of that starting and stopping, and the idle time just sitting around, eats up gas. We found ourselves refueling all along the way, both in the day and at night. So we would top off and then catch up with the assault columns.

There were portions of the invasion where we were on main thoroughfares and we could pick up speed, but then there were also portions across open desert. The amount of combat power out there was amazing. From as far left, far right, behind me, and in front of me, and all the way to the horizon in every direction, I couldn't see a space that wasn't filled with a combat vehicle. It was an amazing sight. I knew what the United States military had, but when you see it all moving across an open desert, and you know that you're only a small little piece of it, and there's bigger shit going on around you, it is … amazing. In the moment, I remember thinking, "This is the focus of the entire world right now." We were in the center of the world stage, and all the eyes of the world were focused on that moment. It was historic.

When the night came, you're only as good as some wad of maps that you had haphazardly thrown at you. I'm talking hundreds of maps, and we would move so fast that by the time we figured out a map sheet, we had already passed that section. Then we couldn't even find the next map sheet. We were just hoping we stayed in the left and right lateral limits and in the general direction that our column was moving.

I had a four-man crew, one of them being a corpsman [*a Navy medic who travels with Marines*], and another being a mechanic, who was the chief mechanic for all twelve of those vehicles in the platoon, Sergeant Scott Dahn. The crew chief was a nineteen-year-old Marine from Georgia, Private First Class Edward Sasser. Normally, the rank for a crew chief is a non-commissioned officer, which would be a corporal or a sergeant. They're normally twenty-two years old, but that was what we had to deal with at the time. So I took the youngest and most capable Marine that we had in the platoon and made him the crew chief of the vehicle I was on. That way I could overcompensate for him. That was still his vehicle, though, and we gave him the respect that's due the crew chief. That is just an unwritten rule in the AmTrac community—the crew chief is the one responsible for that individual vehicle.

Whenever a vehicle would break down along the way, the column did not stop. There weren't any supporting elements that would be coming with flatbed trucks to pick up broken tanks and AmTracs, so we had to get our vehicles running again by ourselves, and then do the best we could to catch up with the column. Sometimes catching up meant forty miles ahead. That's how fast everything was moving. If you had to stop and fix a vehicle, and it even took an hour, you could expect that the column was at least ten miles ahead of you already. Then, all you could do was follow the tracks that they made, get within radio communications range, and then get talked into position and continue on with the forward movement.

That type of speed and action happened all the way until the 22nd of March, which was when we stopped and staged south of Nasiriyah [*a city of about 500,000 along the Euphrates River, about 225 miles southeast of Baghdad*]. We were south of Nasiriyah, but we couldn't see the city. We were staged within striking distance, so as not to alarm anybody in the city that we were coming. We refueled vehicles and reallocated ammunition.

We were up all night on the work team, and then word came: We're going to start moving towards the city at four in the morning. There were armored vehicles as far as the eye could see. When we moved out and got closer to the city, we stopped along the way and let our infantry out to start clearing buildings. Once they cleared a little village or whatever, we'd get them back in. We'd creep up some more until there were some new structures that needed to be cleared, and then we'd do it all over again, all the way to the city.

The city began to wake up and we started to hear fire coming in and out of Nasiriyah. We started seeing our artillery get put into position in the fields, getting set up to fire and cover our movement into the city. We also started to see some aircraft flying over the city, making tactical runs. Our tanks were engaging Iraqi troops in the field that were trying to move. We started seeing more people coming in, hunkering down in the fields and taking pot shots. Then we started hearing mortars and large-caliber munitions coming out of the city, and seeing tracers from anti-aircraft guns. The closer we came to Nasiriyah, the more we got the sense that, "Hey, this isn't some Podunk city."

Ahead of us were a couple of routes: Route 8, which skirts around the western edge of Nasiriyah, and then Route 1, which goes straight in and around the city. There were already columns of Army trucks, supply vehicles, and combat vehicles that were moving through Route 8. They were tailgate-to-nose as far as the eye could see. They were moving way out to the west of Iraq, and they were skirting the city.

Part of that Army convoy was the 507th Maintenance Company, which included Private Jessica Lynch [*Lynch was captured during the battle and her recovery days later by U.S. forces received considerable media attention*]. What had happened to the 507th, unbeknownst to us at the time, was that they fell behind their overall column when they had to do some maintenance. Their column continued to move due to the rapid rate of the plan. So the 507th got separated, and they got lost. Then, that morning, roughly about the time that we were moving up from the positions where we had staged in the night prior, the 507th took that historic wrong turn and instead of moving along Route 8 and skirting Nasiriyah, they went straight into the city by mistake.

The 507th drove through ambush alley with their entire convoy but didn't receive any fire at first. We heard that the Iraqis just watched them and were sort of like, "Is this happening?" So the 507th rolled all the way through the north of the city and then they realized, "Hey, this isn't the route we're supposed to take." Then the shit hit the fan when their captain decided, "Well, we're going to go back the same way we came." And when they did that, the Iraqis were like, "Oh, shit. Here they come again." That's when the Iraqis shot the 507th to pieces.

In my professional opinion, the Battle of Nasiriyah happened the way it did, at the expense of so many Marines, because when the

507th came back through the city and the Iraqis hammered that little Army maintenance column, it gave the Iraqi army, and the entire city, the will to fight. It was like they said, "These Americans aren't anything. Look at what we did to them. Bring it on!" I honestly don't think the Iraqis would have messed with the spearhead of the First Marine Division and Task Force Tarawa if that hadn't happened.

I had about four or five different radio nets in my communications helmet, and I had a box that I could switch back and forth to monitor them all. I started to hear all the stuff going on around me, and it was building an ugly picture, a confusing picture. When we started getting closer to Nasiriyah we heard that our tanks had passed burning Army vehicles. I heard, "Hey, there's U.S. Army vehicles up here, burning. It looks like someone was attacked."

Task Force Tarawa was the strike force heading into Nasiriyah to secure its bridges, and we were told, "Nobody is forward of you." Once those bridges were secure, we were to hold them because the entire First Marine Division piling up in the desert behind us was going to need them. Without those bridges, it was going to complicate matters. We knew what the mission was: to get to those bridges first and hold them for all who would come after us. So no one was supposed to be in front of us, but then we hear about a bunch of burning U.S. Army vehicles up ahead and that confused the crap out of everybody.

Then we passed those Army vehicles, and nobody was in them. There weren't dead bodies lying around. There was nothing there. The vehicles were just empty, bloody, and burning. That was a little creepy. Then we heard a radio call about how one of our tanks thought they saw some U.S. troops off in a field, but the tanks were engaged with Iraqi tanks and moving too fast to stop, and they didn't get a grid position of the troops either. The call came back to me that we needed to take a couple of vehicles and move forward and see what we could do to help. I chose my vehicle and one other, and we drove forward of our own lines. The only ones who were in front of us were a couple of M1A1 Abrams main battle tanks, and they were engaging targets farther out. We couldn't see any soldiers, but I knew that the Marines on those tanks were the last people who did. I told Private First Class Sasser, "Let's drive up to where the front line of the tanks are, pull the vehicles off the side of the road so that they are covered, and then we need to get ahold of the tanks and figure out what they saw."

We went up, parked the vehicles, and I ran up to a main battle tank that was commanded by Major Bill Peeples. They were a reserve unit called Eighth Tanks. I crawled up on the side of the tank, and it startled the shit out of them. I mean, they were hunkered down in their tanks and engaging an Iraqi T-62 main battle tank across that field, and then all of a sudden there was some frigging dude tapping them on the shoulder. [*Laughs*] The shock on his face was like, "What? How the 'F' did you get here"? I pulled up his helmet and said, "Did you see any soldiers?" He said, "Yeah, we passed some about a kilometer back."

I jumped off the side of that tank. I knew enough not to jump straight off the back because that's where the heat signature in the engine compartment comes out. But as soon as my feet jumped off that tank—and he saw me get off—the tank's main gun fired. My feet had not yet hit the deck, and I was hanging in midair when that main gun went off. The overpressure was incredible. It was wild, and very exciting. I can remember the adrenaline. Man, I have never felt that much adrenaline in my entire life ... at least up until that point. The combat power that tank had, the overpressure when that round fired, it sucked my helmet to my head. Then I heard the explosion of the blast. I remember running back to my vehicle with a huge smile on my face.

Those tanks just went on fighting. That's the type of people we had out there. That was our confidence level. It was like, "Hey, this is what we wanted to do. This is what we are trained to do, and now it's game time." Everybody was in the moment. More importantly, everybody was responsible for their own little piece of the battle. You didn't worry about what the other troops were doing. You only worried about what you had to do. That tank had to engage the Iraqi tank, and we had to find some soldiers. And all I knew was that between where that tank position was, and where I had left on our search, there were some soldiers in need of help.

We were in a farm field. So we took our two vehicles and put them on each side of the highway and drove like you would drive down an old country road looking for a possum that you may have hit. We weaved back and forth trying to cover all the areas in the field. After we had been doing that for a while someone said, "Hey, I see something over there." When I looked over the side, I saw a man in the field standing up, waving his arms. I could tell he was a U.S. soldier, so we started heading down the field toward that spot. As we were getting there, the other vehicle called and said they had seen another group of soldiers about two hundred meters away from that first group.

When we rolled up, the scene was, "Thank God you guys arrived." One guy started telling us what had happened, and then he said, "I'm missing half my people. I don't know where they're at." The soldiers he did have were in two little groups. The majority of them had weapons, and they had wagon-wheeled and were all facing outward, defending their position. In the center of the wheel were their casualties. One of them was Specialist James Grubb, who had some gunshot wounds—both legs and in his upper shoulder. He was in pretty bad shape. He was a real big boy, and that's what saved his life, I think.

You won't read about this in any of the history books because the members of the 507th are painted as not being combat troops or not prepared for war, but when I rolled up, what I saw was how their buddies were doing everything possible to make sure the wounded stayed alive. Their medic was treating the casualties better than anybody else could, and they were setting up a defense because they knew the enemy was probably coming back.

When we started loading their casualties in our vehicles, I remember Specialist Grubb smiling and joking around. He looked up at me and said, "I never thought I would be happier than I am right now, to see a United States Marine." But there was another one—I don't know what that soldier's name was, but he was screaming and running around like a stuck pig, almost to the point that you wanted to knock him out to get him in the vehicle. He could still walk, and had some fragmentation wounds, but it was nothing near what Specialist Grubb had. I remember the disparity of it because there was the other soldier who was shot to pieces and was joking around. Positive mental attitude and a sense of humor go a long way, in everything in life.

We got them all into our vehicles and we took them back to our main line. We dropped them off at some ambulances we found, and I didn't ever see them again. As soon as I dropped them off we got the word, "We're now going straight into Nasiriyah." There was a sinking feeling in my stomach, because up until that point it was all just the pregame. The big show was about to happen. We started loading the infantry and staged in our battle columns. The final orders went out, and we started moving toward the city. I can distinctly remember there was a railroad track and a railroad station along the entrance into the city. When we crossed that railroad track, shit got real ... and real quick.

When our lead vehicle crossed that railroad, we saw a white van with a blue stripe that was driving at a high rate of speed towards the column. It was coming head-on. Then it turned to its side—that's how I can remember seeing the blue stripe. I heard somebody yell, "Rocket!" You could see a gunner with an RPG [*rocket-propelled grenade*] hanging out the window. We yelled over the radio net, "Sagger! Sagger! Sagger!" which was the rehearsed code word for a specific movement when under attack by a rocket. As soon as we said the code, all our vehicles started driving erratically, shaking their armored ass-end, for a lack of a better description, shaking them back and forth, and zigzagging like a destroyer would while trying to get away from a submarine in World War II. The RPG gunner missed our vehicles. As soon as that van was out of sight, I didn't even have to say anything. Our vehicles instantly fell back into their assault columns and kept moving forward. That was the point when I knew, "We're going to be okay." The instantaneous response of those young men when they heard the code word "Sagger" gave me confidence that they'd stick to their training.

We had done a portion of that training back home. We were trying to save gas so I took a few of them to a golf course at Camp Lejeune and grabbed a bunch of golf carts. We drove the carts not to play golf, but to train for war. We drove them in simulated armored columns and practiced and rehearsed those kinds of drills. Then on that ship we would rehearse all of those code words. We would rehearse what this word meant, and what that word meant, because we knew that it could save lives during battle.

So, after the RPG attacked, we instantly re-formed into the assault columns and headed for the southern bridge over the Euphrates River, which was our objective. It was eerily quiet along the sides of the roads. When we were rolling in we saw some Iraqi tanks burning. Our Combined Anti-Armor Teams, or what we call the CAAT platoons, are Marines who are mounted in Humvees with TOW missile systems [*tube-launched, optically tracked, wire-guided missiles that are commonly used against tanks*]. The CAAT platoons are usually the eyes and ears out in front of our tanks patrolling and identifying targets. They had already taken shots and killed numerous tanks along the way. They were all staged prior to that bridge, waiting for us to come through and secure it.

I can remember rolling up on that southern bridge, and for that brief moment, you couldn't hear gunfire; you couldn't hear anything. We sat there for a few minutes. The bridge spans probably a good couple hundred yards over the Euphrates, and it was long enough that you could take an armored column from one end of the bridge to the other and fit all twelve AmTracs. I yelled to my first section leader over the radio, "Hey, you guys need to push up farther because I'm sitting exposed on top of this bridge." I was the last vehicle, because the platoon sergeant is always last. He makes sure everything else is taken care of, brings up the rear so nobody gets left behind, and then watches everyone's back. When we crossed the bridge we off-loaded the infantry because our job was then to secure the bridge.

I can distinctly remember as soon as we moved across the bridge and into the intersection, the entire world blew up. And when I say "blew up," that's not an exaggeration. Numerous rockets flew through that intersection at the same time. Artillery from somewhere within the city was firing and dropping rounds onto our position. There were muzzle flashes coming out of every single window that could be seen. I remember saying to myself, "This is what Custer must have felt like at the Battle of the Little Bighorn," because I could not turn in any direction on the face of a clock without seeing rockets or muzzle flashes, from alleys, from buildings, from everywhere.

We quickly put the vehicles in their battle positions and blocking positions, covering numerous alleyways, covering the roads, and covering the bridge. Looking to the rear, I could see people coming out from underneath the bridge with rockets and guns who were firing behind us. I knew this was pretty well rehearsed, probably the official defense of the city. They had planned who was covering what positions and what could happen when the Americans would have to come into the city and cross that specific bridge. They would have to fight. People started screaming over the radios. I started hearing Marines getting hit, and people calling for a medevac. This all happened in the flash of a few minutes.

Since the firepower was so overwhelming, did your platoon ever think of retreating from the bridge and regrouping somewhere else?

No. Marines get beaten into them, for lack of a better term, that the number one thing is mission accomplishment. We never retreat. We always move forward. The second thing we get beat into us is troop welfare. I also believe, as an ethos of Marines, Marines understand their significance in history. They live it. They are taught it. It is reinforced into them. They know their battles. They live all three verses of the *Marines' Hymn*, the first line, "From the halls of Montezuma to the shores of Tripoli," which discusses two different battles in Marine Corps history. The last line, "If the Army and the Navy ever look on heaven's scenes, they will find the streets are guarded by United States Marines." That's from a battle called Chapultepec in Mexico City in 1847, in which we stormed the Mexican Military Academy, against all odds. When General Winfield Scott rolled in there with the U.S. Army, he didn't expect to see anything but dead Marines. Instead, he found battered and bloodied Marines lining the streets and proudly standing at the entrance. Not just Justin LeHew knows that; every Marine knows that. So we understand our significance in history. We understand who came before us and the challenges that they faced, and more importantly, what they had to overcome. It's put into us that Marines always do more with less, and the last thing the United States should ever expect is the failure of a good Marine.

There's a saying: "We breed them differently." We breed an expectation that every Marine understands what they're responsible for, regardless of race, creed, or color. You don't care who the president is. You don't care why we're there, because you're in it, and if you don't do it, you're going to die. It's pretty easy to make decisions when you're surrounded by Marines like that. During their training we also exposed our young Marines to stories about the men who survived being captured in North Vietnam. I showed them the amount of torture and the amount of pain those men endured so they would have an understanding of what might happen. Collectively, they bonded together to make sure they weren't ever in that position.

That's the fighting force that held that bridge in Nasiriyah. They protected each other. They protected their brothers. I looked off the side of my vehicle once and saw two young infantrymen sitting back to back, covering an avenue of approach, and they were handing ammunition magazines back and forth to each other. They were both probably about nineteen years old. I was so damn proud of them, because in a short amount of time we trained them, and put the values into them that allowed them to be doing what they did. That is what's most amazing about the training and the transcendence of the U.S. Marines; we took thirteen weeks of boot camp plus four weeks of infantry training, and we have that.

No corporation or anything in existence can inspire that much in their people to perform to that same level. Nothing can match that. Those are the ones that hold the line. Those are the ones that write the history of the Marine Corps. It isn't the generals. It's the stories of those men that don't get told, and are never going to hit the history books, that make up this great nation that we are in. It's that average American citizen that shows up and says, "Man, on my watch, that shit isn't going to happen." I was lucky enough to serve with hundreds of such men.

On the bridge, you said you saw Iraqis coming at you from underneath the structure?

Yeah, they were coming out from under the bridge. Then they were behind us. It was all pre-planned. There was firing coming from the rooftops, out of the windows. It wasn't concentrated fire per se, like they knew what they were doing. It was kind of "get up in the window, spray a weapon, and then get out of the window." It was just a large volume of fire being thrown out into that intersection. I thought, "I'm inside an armored vehicle, but we have infantrymen outside who are in a 360-degree firefight." I knew we could defend ourselves. But normally we would concentrate fire into a certain portion of a perimeter. Here, the perimeter was 360 degrees with targets everywhere. We hadn't trained for that type of warfare. So we were trying to react to everything. Marines were trying to yell over the large volume of fire. They were trying to fix fighting positions and identify targets. Civilians were in the midst of it all, too. I looked down an alley and saw a crowd of about two hundred to three hundred people gathered to watch, so we couldn't just fire away at anything we saw. We really had to pick and choose our targets, and the Marines train better than anybody to do that. We're trained to have "positive identification" before pulling the trigger. The Marines started firing in front just to keep those crowds back. All the while, other Marines were still dealing with targets on the rooftops and dealing with rocket fire.

The commander of Alpha Company was Captain Michael Brooks, who was a phenomenal commander. I would not have wanted any other commander in Nasiriyah with me on that day. Captain Brooks was parked right behind my vehicle talking to me when a rocket trail streamed overtop of the vehicle and exploded about ten feet off the front side. He just looked up at me and said, "What the hell was that?" I told him, "That's an RPG." The Iraqi RPG gunners would get on top of rooftops, Marines would start shooting at them, and the gunners would fire their RPGs erratically. Some of them would hit power lines. Some would skip off the deck [*Marine-speak for the street, the floor, etc.*] and then explode off to the side. But of all twelve vehicles I had in that battle position, two of them got hit with RPGs, but both times they glanced off the side or front and never exploded.

As soon as that RPG went over my head, that was a cue for my tractors to start moving. One of the reasons why I think that our vehicles didn't get their butts handed to them that day was because we made sure that they never stayed in the same position for more than a few minutes. They would move. One would cross the street while the other would go to a different corner, just so the Iraqis couldn't simply fix their sights on a large vehicle sitting in the middle of an intersection.

Our attack was still very coordinated. The infantry were covering behind any kind of wall or a little pile of dirt, so our vehicles had to look out for those guys when they were moving around. On the other hand, the infantry had to look out for our vehicles at the same time they were looking for the enemy on the rooftops. It was a miracle nobody got run over during all of that movement.

We took up a battle position covering an alleyway—my vehicle, Alpha 312, and a second, Alpha 311, crewed by Sergeant Joshua Collins. I could see down the alleyway probably a good three hundred meters. All of a sudden, an ambulance started coming at us at a high rate of speed. I knew it was an Iraqi ambulance because we had studied photographs of many types of vehicles before the invasion. It had a flashing light, too. It was as large as a bread truck and on one side it had a Red Crescent moon—the Arabic symbol for an ambulance. Anyway, it started coming at us a high rate of speed, maybe sixty kilometers per hour. That was a highly unusual speed to be driving down an alleyway, and it made me think they were about to attack our infantrymen. We fired warning shots, but the vehicle didn't stop. I needed to make a call, and make it quickly.

What were you thinking?

I grew up in the generation that remembers the truck bombing that killed the Marines in Beirut. That was just a regular U-Haul-type truck, and I remembered how they used it to mask a truck bomb. We had trained that nothing got within our perimeter of seventy-five meters. As the battle kept going, unfortunately, you'd let them come a little closer so you could make as much positive identification as you could before engaging. Secondly, we'd let them get a little closer just to give them a chance to get the hell away once they found out what types of weapon systems they were facing.

But that ambulance didn't stop. So I aimed our .50-caliber right into the cab and laid in a ten-round burst of fire. The cab splattered red, and the ambulance veered off the side of the road and hit a telephone pole. When it stopped, the back end of that ambulance opened up and out came six uniformed Saddam Fedayeen soldiers in black uniforms with red triangles on their shoulders, all carrying weapons. That's when we all started thinking, "Hey, wait a minute. They're going to use ambulances. They're going to hide in the civilian population. This really isn't the kind of warfare that we prepared for. How do we deal with this?"

What LeHew and his Marines were seeing was happening all across Iraq. "Iraqi forces violated international humanitarian law during the ground war, directly causing or contributing to civilian casualties," according to a report by Human Rights Watch. The group's field investigators found that Red Cross and Red Crescent emblems were being used on military vehicles, civilians were being forced to become human shields, Iraqi soldiers were fighting from hospitals and mosques, and soldiers were wearing civilian clothes. In one instance, Human Rights Watch documented how an insurgent hid behind two women, one who was carrying a child, and another time when enemy fighters lined up civilians in front of their vehicles so they could advance without risking U.S. fire.

Back in Nasiriyah, the Marines pressed on, respecting the laws of war but never allowing the enemy to gain an advantage.

Sergeant Collins and I did what we had to do in that situation, and we adapted. We followed the rules of engagement that we were taught, and we each had our own personal conviction: "Hey, I'm going to identify a target to the best of my capability, and if I consider that target a threat, then I'm going to pull the trigger." That's exactly what the Marines out there had to do, to make those assessments because communications were sporadic at best. Those Marines needed to make life-and-death decisions instantly, without getting on a radio asking leadership, "Can I take the shot? Can I do this?"

That proved it. Once that ambulance was hit, it was total warfare inside of that city. There was a guy named Gunnery Sergeant Merryman, who was basically in charge of supply, logistics, and everything to support the infantrymen who were out there. He was a sniper by trade. A young Marine had run up near my tractor and I heard him talking to Gunny Merryman. He said, "Gunnery Sergeant, can you come and look at this?"

He explained how about one hundred yards down an alleyway, there was a woman who would come out of this building every so often, and she was holding something. Then, after she went back into the building, an RPG would be shot from the top of the same building. They had watched her do this four separate times. Gunny, I, and everybody else knew she was a spotter for the RPG team on that rooftop, calling out targets. Now, it's simply not the American way to engage civilian women. The gunnery sergeant and the young Marine sat and watched her come out one more time, and then another RPG was fired. When she came out the next time, that Marine—he was a distinguished marksman—dropped her right in the forehead. We didn't receive any more RPG fire from the top of that building. Marines then started moving down and started to clear structures in that alleyway because of the hard decision that was made.

When we looked on top and around the corners in the buildings, people and families would actually come out onto their verandas and watch the battle, which was the most eerie thing that I can remember. Looking up and seeing people sitting there, as if they were at a cafe stand sipping chai tea. It was strange. I will tell you, for the most part, I think we figured out that they weren't just there to watch the battle, either, as spectators. Some of them were acting a little too normal. In the opinion of a lot of Marines, some of those people were probably the commanders. Some of those people were probably the lieutenants, and many were reporting on our positions.

We were in that position for probably about an hour when Private

First Class Sasser came over the radio and said, "Hey, look at those dumbasses." I turned around in the turret and saw an AmTrac going the wrong way. Like it was going through our battle position back over the Euphrates River Bridge. It looked like somebody was scared and they were making a run for it. I remember thinking, "God, don't let that be one of mine." As soon as I looked at that vehicle, from the top of a hotel an Iraqi gunner shot an RPG that went straight into the AmTrac's top cargo hatch and exploded inside the vehicle. It blew the back of the ramp open. Then, out of the corner of my right eye, I saw an Iraqi gunner run out about fifteen meters into the middle of the street, drop to one knee, and shoot an RPG into the open troop compartment of the disabled AmTrac. That all happened in seconds—from the first shot in the top of the vehicle, to the ramp being blown open, to that RPG gunner coming out and firing a shot directly inside the troop compartment.

The AmTrac came to a rolling halt, and I saw a Marine fall out of the back ... he was on fire. I was thinking this was kind of like a "Black Hawk Down" situation because that was a catastrophic kill on the vehicle. I didn't think there would be any survivors, and that's a vehicle of the U.S. Marine Corps. It's got weapons. It's got classified material. It's got our radios. It would be a game-changer if it fell into enemy hands.

It wasn't one of the AmTracs in your Alpha Company, right?

No, it wasn't. It was from Charlie Company, a unit whose mission was to secure the northern bridge over the Saddam Canal, but they went through what is now known as Ambush Alley. They crossed that northern bridge, and then unknowingly arrived at a pre-plotted Iraqi artillery position. So the Iraqis started raining down artillery on the Marines.

Meanwhile, Bravo Company was supposed to secure the outskirts on the right-hand side of the city while Charlie Company was supposed to secure Saddam Canal on the northern bridge, and then keep that roadway open for the First Marine Division and everybody else who was heading to Baghdad. But Bravo Company found themselves on top of a sewage canal, and the ground wasn't stable enough to hold armored vehicles. The vehicles started sinking in the mud, and then the Iraqis started to attack.

Bravo Company had a forward air controller that was with them, providing information to two U.S. Air Force A-10 Warthogs, which are tank-killers. So the forward air controller cleared the A-10s to fire at anything north of the northern bridge, because nobody was supposed to be that far north. The pilots saw Charlie Company's vehicles, but weren't used to working with Marines so they mistook them as being Iraqi vehicles full of fighters heading to reinforce the city.

It's called a "friendly fire" incident. The depleted uranium 30-millimeter rounds from the A-10s chewed through Charlie Company's AmTracs, and there were also a couple of missiles that were fired. It completely leveled a couple of Marine AmTracs. So all of a sudden their entire command and control was broken. They have Iraqi artillery firing at them, and they also have their own United States Air Force chewing them up. At one point, a young lance corporal grabbed an American flag and ran with it down the center of the road, waving off the last attack run that was coming in.

That's still really questionable. When those A-10s went over the top as they finished their runs, Marines said they could see the bolts on the bottom of the planes. For somebody to be a couple hundred feet off the deck by the time they finished their gun runs, and to not realize what they just did, and to circle back around and make two or three runs ... I think that was the biggest psychologically damaging thing to happen to those Marines. They couldn't figure out what was going on. So then they jumped in their vehicles to medevac their wounded Marines. To do that they had to put them in the back of their AmTracs and run those vehicles all the way back through Ambush Alley again, back through our battle position, then finally outside of the city to get medical help.

As they started making their medevac run, the Iraqi RPG gunners started picking them off. Littered all the way along Ambush Alley were five or six other burning AmTracs that didn't even make it as far as my position.

Back to the one that got attacked on my bridge ... soon after that devastating RPG attack, a second AmTrac from the same unit arrived. Its crew chief got out and kept yelling, "You gotta come with me to help out my friends." But the Marines kept him right there. I looked, and nobody was running towards the AmTrac that got hit. Another eerie sight. Just a burning U.S. vehicle in the center of the road, and nobody was going towards it because, quite frankly, they had their own stuff they were dealing with and they were probably in shock and disbelief at what they just saw, and they probably figured no one survived.

We were still in that cluster situation where small arms fire and RPGS were coming at us from every angle. I looked in the back of the burning AmTrac and I told my maintenance chief, Sergeant Dahn, to get up in the turret and cover my job. Then I grabbed a nineteen-year-old corpsman named Hospitalman Alex Velasquez and told him to grab his unit one kit, which is the combat medical bag, and said, "Come with me."

I took my communications helmet off, but I didn't put another helmet on. At that time we were still inside of what we call MOPP levels [*Mission Oriented Protective Posture*], which is all of our chemical-weapons protection, basically heavy outer garments, and it was already about 130 degrees out.

So you're fighting the Iraqis and the environment?

Yes. Marines everywhere were dehydrated. They were out in the street, with corpsmen putting IVs into people while they're still fighting, just trying to keep people from dehydrating. There were no supplies coming in, and the Marines were sucking down water like crazy. We had to slow down the water intake as well, because we didn't know when supplies were going to come up.

I jumped out of the back of my vehicle, grabbed the corpsman, and we ran over to that burning AmTrac. When we got there, the first thing that I saw was half of a leg of a U.S. Marine laying on the vehicle's ramp. It was blown off from the knee down ... still had the boot on. I picked it up, handed it to Velasquez, and said calmly, "We're going to find who that belonged to." Velasquez laid it off to the side, and we went in the back of that vehicle.

Was it still burning?

Yes. The vehicle was on fire. There are three cargo hatches on an AmTrac. One on each side of the top that opens up to allow viewing, and then there's a center-beam cargo hatch that normally stays closed. The only time you ever open that one is if you're going to load cargo down into the top of the AmTrac with a crane. It always stays closed. But there weren't any cargo hatches on the thing. They exploded and were laying down inside the vehicle, broken, on top of the Marines that were still inside.

There was no movement at all. The corpsman, who we called "Doc," and I went through and started grabbing Marines. The first one I grabbed ... when I pulled on his upper body it separated from his lower body. He was dead, of course. I then knew it was a Marine by the name of Sergeant Mike Bitz.

The Associated Press reported his death a few days later: "Sgt. Michael E. Bitz, 31, died without ever having seen his youngest children, twins born one month ago after he was dispatched to the Persian Gulf ... according to his mother-in-law, the father of four was 'a damn good dad.'" In addition to the twins, Bitz also had two sons, Christian, 7, and Joshua, 2."

I knew Mike, and I can still see his face. I looked at Doc, standing there holding the upper torso of a U.S. Marine. Then a Humvee pulled up outside. I don't know if it was Gunnery Sergeant Merryman or who it was, but they started loading up the body pieces inside the Humvee. Doc and I continued to triage and go to anybody we could find. I was in front of Doc by about five feet, trying to pick wreckage off Marines while Doc felt for pulses.

Fuel was everywhere because the blast had cracked open the fuel cell. Smoke was everywhere, too. It was a sickening smell. There were still six hundred grenades and about ten thousand rounds of .50-caliber ammunition that was in the deck plates of the AmTrac. It still had its weapons systems, and it still had its communications gear. I looked at Doc as he was feeling for a pulse on a Marine, but from my position I could see that the Marine was missing half his head. Doc couldn't see that from where he was, though, so I said, "Doc, let that one go. Go to the next one." That happened a few times.

We had twelve or thirteen Marines who were in the back of that thing. They were all crushed underneath that material, and there was no movement at all. I figured they were all dead. The vehicle had weapons, though, and I knew we needed the weapons and ammunition because it was going to be a long day of battle. I went up to see if the weapons were still good. I also knew the vehicle was going nowhere and we were going to have to either take the radios or destroy the radios, so I planned to either melt them with a thermite grenade or pull them out of the vehicle and destroy them, smash them. It's one of those decisions you didn't want to make but the simple fact was, well, there were dead Marines inside of it, and there is still a battle raging outside. We had to do what we had to do to stay alive, and stay in the fight.

As I was stepping forward inside the vehicle, I heard a Marine gasp for air. We started peeling back casualties, and underneath one of those collapsed cargo hatches was a Marine named Corporal Matthew Juska. I looked down at him, and he was bent over in an L-shaped position. The cargo hatch had crushed his back, and his head was split all the way from his neck to the top. I really didn't think he was going to make it. We started pulling at everything, because at that time the idea is we had to get him out, not worrying about whatever injuries he might get by moving him. We had to get him out of there, or else he was going to die, regardless.

Somebody outside started yelling, "Hey, get some people in there to help." Then I heard someone else yell, "Screw them, they're all going to die in there." That's not something I thought I'd ever hear.

I looked at that corpsman and said, "Hey, Doc, get out of here." I was going to stay.

I knew we were sitting on top of a time bomb, with all the fuel, the fire, and the unexploded ordnance. That sailor from Puerto Rico looked at me and he said, "I'm not going anywhere as long as you're here, Gunny." Absolutely amazing.

So we continued to dig that Marine out. It was hard. I found out later that he was one of the biggest guys in their company. He was 6'2" or 6'3", maybe taller, and weighed over 200 pounds. Then he had all his battle gear on, too, so he easily weighed close to 300 pounds. To this day, I swear, him being so big probably saved his life. We eventually got him dislodged enough to where, with the help of a couple of more Marines, we ran him back to my tractor and Doc took care of him in the back of our vehicle. I climbed back up in the turret and relayed to Lieutenant Brenize that we had a casualty who needed to get out of there now.

They were originally calling for an air medevac. I didn't hear anything about the air, and just about the time we were going to decide whether to run him out of there via ground, we heard there were other wounded Marines in a house nearby. I told the lieutenant that I was going to find where that was.

I remember seeing weapons laying all over the place. So you didn't have to really worry about where your weapon was; you just picked up whichever one was on the street. I went over to some Marines along a wall and asked them if they knew where there were some casualties in a house. They pointed about five houses down, and said, "Gunny, I think they're down that way." Then they went back to shooting. They were defending one of those alleyways. They were scared, and I looked back at them and said, "Hey, man, I bet this wasn't what you had in mind when you went into the recruiting office, was it?" They started laughing. I started laughing. They went back to their work, and I ran to the next battle position.

I ran over to a house, went inside, and saw two Marines slumped against the wall. I could tell they were AmTrac crewmen. They were part of the crew from the vehicle that had been hit with the RPG. They were blown out of the hatches when the RPG hit. They were dazed and confused, gray and ashy from head to toe, and they had blood coming from their eyes, their noses, and from their ears. They each had a glassed-over look, but they were still alive. I could tell they were completely concussed, as well. They had no idea where they were. I saw another crewman who was moving. I said, "Hey, man, I'm going to get you out of here." That Marine said, "Gunnery Sergeant, I can still fight." I looked down and he was missing the back of his foot. He was missing a chunk out of his leg, too.

I could hear Iraqis in the back of the house. I didn't know whether they were a family, or whether they were combatants. I could hear people yelling. You could hear doors opening and closing. Someone was in the house and there was nobody to defend the wounded Marines if I left. So after I patched him up to stop the bleeding, I racked a round in an M16, handed it to the wounded Marine, then put him in the doorway. I said to him, "I'm going to go get help outside. You stay in the doorway here. If anybody comes from my direction, do not shoot them because I'm bringing Marines. If anybody comes from the other side of that house, then do what you gotta do."

I was gone just for a few moments. I ran outside, gathered up some Marines, and when I ran back outside of the house, Doc started scream-ing at me from the AmTrac across the street, "We gotta get this guy out of here. He's going to die." Then I heard Lieutenant Matt Martin, who was the executive officer of Alpha Company, yell, "We've got a helicop-ter coming in." We started moving the Marines out of the house and consolidated them against the wall near the AmTracs.

Meanwhile, the firefight was still going like crazy. You could just see the tired looks on the faces of those Marines. We were probably two and a half hours through nonstop fighting. Not to overstate it, but the type of fire that was coming at us was like the fire that the German infantry put on the first wave at Omaha Beach. It was every-where. It was sustained, and for hours. There was no lull in it, no break.

Major Tuggle, the executive officer of the battalion, asked if any-body had smoke because we had helos coming in. There was a CH-46 Sea Knight circling above [*a large, tandem-rotor helicopter that's about forty-five feet long with rotor diameters of about fifty feet*], and I remember thinking, "I have no idea how that is going to get inside of here." It was still a hot LZ [*landing zone*], there were power lines and poles all around, and it was an intersection.

At first they gave the officer a pyrotechnic to signal the pilots, and this pyrotechnic looked like a little handheld rocket. He was standing in the street, fired that rocket, and it skipped off the pavement, shot somewhere erratically and then disappeared. The major was standing in the middle of this firefight, looking at this rocket disappear down the street. He puts his hands up in the air with his empty rocket tube and starts laughing. So I started laughing at him. In situations like that, you try and keep people's heads in the game. A little bit of humanity and humor helps, believe it or not. Somebody ran out to the officer and handed him what looked like a smoke grenade. It was purple smoke. He pulled out the pin and threw it. The purple smoke started billowing, and that was going to be where the helicopter came in. As soon as that smoke went off, we ran back to the vehicles and started getting the casualties ready.

We started running towards the purple smoke. The CH-46 came in from the sky and went towards the smoke. We didn't know that purple smoke to a pilot meant, "Don't land here. This is a hot area." As soon as he saw the purple smoke, he stopped and hovered about fifty feet above the deck. As we were running casualties out, the helicopter slid sideways over the top of our head ... he was going back toward the intersection where the AmTrac got hit. Somebody over there had thrown some green smoke, which to the pilot meant that was a clear place to land. But the intersection was absolutely not the place we wanted him to land. Green smoke to the infantry meant, "Shift your fires away from here," but to that pilot it meant, "Land your helicopter there." We gathered the casualties quickly, and then ran them back another hundred meters the other way. A guy named Major Eric Garcia put that helicopter down in the center of that city, dropped the ramp, and we ran casualties to the back of the CH-46.

When we started loading Corporal Juska, we got into an argument with the crew chief. He had the litters inside and wanted the casualties to be neatly strapped in ... all stacked up nice and tight.

I told him, "Hey, man, you don't understand what's going on out here. You need to get this helicopter out of here, right now."

He kept arguing, wanting it done a certain way.

I said, "Do you know what an RPG is?"

He said, "RPGs? Are they firing RPGs?"

I remember saying, "I don't know if they have any left, they've fired so many."

He just said, "Drop him. We'll deal with it."

No sooner were my feet off the back of that ramp than that helicopter was lifting off. I ran over and dropped near a wall with two other Marines. We could barely catch our breath. The last thing that I saw of Corporal Juska at that time were his legs hanging out of the back of that helicopter with blood dripping from his boots as it flew off, and then that crew chief lying over on top of his body holding him inside the helicopter. I can remember distinctly thinking, "Where do you get guys like that?" That individual was going to hold him inside the helicopter with his own weight.

He didn't know who we were. He was just doing his mission. They knew Marines were down there. They put their lives on the line. They came in to the hottest position, and I damn sure know that they've studied "Black Hawk Down" and know about what Warrant Officer Michael Durant went through. The crew chief might not have known, but the pilots? I think they knew what they were flying into. I always carried a notebook with me and would try to write down things that I saw Marines doing, even if it was just a couple of sentences. It was just a reminder not to forget about stuff so when I eventually had time, I would write citations for valor. Right there, while we were catching our breath against a wall, I wrote about a paragraph recommending the Distinguished Flying Cross for those pilots.

Right after that, someone threw a canteen to us. It was half full, and between three Marines we sucked it down. That was also probably the closest I ever felt to being completely dehydrated in my life. I remember seeing tunnel vision, and after that water it kind of went away. If it weren't for that little bit of water in that canteen, I might have been a casualty laying down there that day.

As soon as that helicopter left, Captain Brooks came up and said, "Gunnery Sergeant, we are going up to reinforce that position on the northern bridge."

Just then the tanks started showing up. I distinctly remember the cheer of our Marines when those tanks came over the Euphrates River Bridge. When they started engaging targets, it gave everyone who was completely worn out an extra boost. At one point I went over and picked up the phone that hangs on the outside of the tank. It's used to talk with the crew inside. I told one of the tankers, "Hey, man. There's a building across the street. It's red, and it's got these windows up on the top level. Knock off the second level of that building because that's where we're taking sniper fire from." Within seconds the entire top of that building was gone. When the tanks showed up, that's when the Iraqi fire started calming down because they knew, "Hey, this isn't that little Army unit that we dealt with this morning. These guys have tanks. These guys are huge."

We mounted everybody up, and I took the last vehicle in the entire column. We put two tanks in the front, and I told the AmTracers, "Train your gun turrets on the tops of the buildings." We had the cargo hatches exposed. Everything was opened because they normally carry seventeen or so combat-loaded troops, but that day we were putting up to twenty-three. We were putting people on top. Everybody got a ride.

So the AmTrac guns were trained on the rooftops. The tanks trained their turrets onto the other side of the road, and we went very rapidly up Ambush Alley to the northern bridge. That's when we started seeing how bad everything was for the other Marines. We saw AmTracs on fire all through Ambush Alley. We saw Marines, and pieces of Marines. When we got up to the northern bridge, that's when reality hit everybody. The carnage was unbelievable.

A Marine walked up to our column, and he kept screaming, over and over, "Did you see what they did to us? Did you see what they did to us?" Then I saw another Marine sitting on the side of the road with his head in his hands ... crying. He was almost unrecoverable, totally inconsolable. Eventually, in order to snap him out of that, I had to take him over to where his boys were wounded, and have him sit with them, talk to them.

I said, "Hey, man, these Marines need you. You're their leader."

That's what you do. You don't pass judgment. You hope someone would do that for you. You just put yourself in a position to say, "Hey, man, we're seven thousand miles away from home, and all we've got is each other. Here's what we need." It snapped him right out of it, and then he started taking care of his boys, started doing what he had to do, and performing just like I knew he would. Sometimes we all just need a little bit of a kick-start, you know?

That's when the Iraqis started backing off the northern bridge. When they saw us, they knew something bigger had arrived. We started setting up a defense, resupplying what we could across the lines, getting heads and tails made up of casualty counts, who's missing what, what kind of gear is needed. All that. The following days were pretty bad, too. All of my Marines from Alpha Company were accounted for, but there were Marines from other companies who were missing. And we already knew that the Iraqis had captured some soldiers from the 507th Maintenance Company, so we all spent a few days searching street-to-street, and house-to-house.

A few days later someone said, "We got someone coming over the bridge." It was a female Iraqi. We had snipers covering that bridge, and they were almost going to shoot. But then we found a lance corporal who could speak broken Arabic and put him on a bullhorn. He got her to stop halfway on the bridge. Then we sent Marines to check it out. I can remember them bringing her in, and in her arms she held the bloodied dog tags and the MOPP suit from a Marine named Lance Corporal Michael Williams.

She then told her story. Turns out that Lance Corporal Williams was one of the Marines who had been wounded in Ambush Alley and was inside one of those AmTracs that had gotten shot up on its way trying to medevac him and others out of the city. After we had left those battle positions, she had gone out, dragged the lance corporal's body into her backyard, dug a shallow grave and buried him. She did that so the Fedayeen fighters wouldn't find his body and exploit it, abuse it. So after we heard her story, she took the Marines to her backyard and they uncovered and recovered Lance Corporal Williams' body.

A few days later a newspaper back home carried this story: "'Mike is gone.'
That was the note his mother sent early Saturday morning. Mike Wil-
liams, a 31-year-old Marine lance corporal from Phoenix, was killed

*March 23 in the fighting near Nasiriyah, Iraq ... His family was told
early Saturday that he was dead, just three days after learning he was
missing in action. 'We have a deep faith in God,' said Williams' mother,
Sandy Watson. 'We know Mike is with him now.'" His brother Joe, also a
former Marine, told the reporter that his brother had 'died a hero ... He
died fighting for his country.'"*

That Iraqi woman had the wherewithal to protect a United States
Marine's body who, by some perspectives, was an invader in her
country. But she shielded his body and made sure it wasn't thrown
around on television.

*Elsewhere in Iraq the reprisals and suicide bombings that would mark the
coming years were just beginning to show themselves. Regime loyalists
targeted anyone who displayed submission, or even hospitality, to coalition
forces. A woman was "waving a white flag to get out of an area that was
hazardous," said Major General Victor Renuart of U.S. Central Command's headquarters, in a New York Post article. "Our troops allowed her
to continue. They continued on a patrol, came back sometime later in the
morning and found her hanged at the light post on a street corner." Regime
loyalists killed the woman in order to send a message to other Iraqis:
cooperation with coalition forces will get you killed.*

*Five days later, Renuart's command released the details of the second
suicide bomber to strike coalition forces since the invasion. It was an attack
on a military checkpoint northwest of Baghdad by two women, one of
them pregnant. "A pregnant female stepped out of the vehicle and began
screaming in fear," the military's statement read. "At this point the civilian
vehicle exploded, killing three coalition force members who were approach-
ing the vehicle and wounding two others." After the attack, Arabic
television stations aired pre-recorded statements from both women, who
spoke of "martyrdom" and war against the "infidels."*

About the end of March, maybe the 30th, we were at a battle position
near the western intersection. It was late at night, and Captain Brooks
called me over to a vehicle. Standing there was an Army lieutenant colonel

and another soldier. They were Rangers. He said, "Are you Gunnery Sergeant LeHew?" I said, "Yes, sir, I am. What can I do for you?"

He said, "Well, we'd like to say it's more about what we can do for you. We have it on good authority that your missing Marines, and our missing soldiers are buried in a soccer field behind a hospital downtown. We're going to go get them." Then he said, "We heard what you did for our soldiers on the other side of the city, and we just want to thank you, and tell you that we will stay as long as it takes to find your Marines."

I remember how good that felt, for an Army Ranger, who wasn't co-located with us or anything else, to come up and say that. The Rangers inserted into the soccer field that night, and a Special Forces team went floor-by-floor in the hospital and found Private Jessica Lynch. While this was happening, our unit was in charge of a diversion to draw the attention of the Iraqis away from the hospital. Those Army Rangers dug with their entrenching tools by hand for two hours in that soccer field, and they found every missing American, both the ones that were captured from the 507th Maintenance Company and the missing Marines. We finally had 100 percent accountability.

After that was done, we just started holding our positions and kept fighting while the larger divisions started pushing to Baghdad. We kept fighting around the city for another two weeks until we dislodged the enemy. After that, we got back on the road, caught up with everybody else, and went into different cities. Fighting was sporadic during the rest of the deployment, but it was never again anywhere close to the intensity of Nasiriyah.

Whatever happened to that Marine you pulled from the back of the burning AmTrac?

When we got back to Camp Lejeune, the First Sergeant, First Sergeant Thompson—a great Marine, he was—said, "Hey, Gunny, we've got something to show you over here." When I went in the back of the building, Corporal Juska was there. There weren't any words ... just a big hug. I was really happy to see that Marine alive.

Looking back, I take solace in the fact that we trained our Marines very well, and that they could also fight morally and ethically. I was a ravenous student of warfare, and I had studied the effects of troops coming back from combat. When you go to war, the decisions you make are going to affect you for the rest of your life. So you don't want to compromise yourself out of a fit of rage because some terrorists took down the Twin Towers. You don't want to shoot up a family because you're pissed off and because you think they have weapons. That's not going to affect you in the next twelve months, but it will affect you twenty or thirty years from now when you have your own families. So my goal was to make sure that that kind of mentality went into the training and preparation phase, so that when they were in the action phase the decisions that they made were decisions they could live with for the rest of their lives.

As far as PTSD goes, I don't really have it. At least I don't think I do, which some doctor would probably say is a classic symptom of someone who has the worst form of PTSD. Of all the situations that I or my men were in while in combat, the decisions we made, they were right, they were just, and they were decisions that I knew I could live with at the tail end of the war. I can look at my grandson and be very proud of what his granddad did in Iraq. I don't have to worry about looking at my family after throwing machine gun rounds into somebody's house when it wasn't justified.

Ninety percent of my Marines have assimilated back into society, are back with their families, and are no longer serving in uniform. They are great husbands, fathers, and sons, and they can look at themselves in the mirror and say, "Whether I believed in that or not, I ponied up for the cause. I did what my country told me to do, and I did it in a justifiable manner, regardless of what the public sentiment was." They can say, "I had a mission, and I went and did that mission. I kept it within the left and right lateral limits and I came home. More importantly, I came home with my honor and my dignity." That legacy resonates with me more than any combat heroics.

I get a great feeling watching those men talk to each other on Facebook, which is the weirdest thing I have ever seen. It's great to watch the photos they put up with their families, and watch the little reunions they all have every year. To know that the training that we did, both in mind, body, and spirit, helped put them in a position where they could be the best that they can be, not only for the service that they gave in the war, but also to be better citizens when they came home. That is what keeps the nightmares away, knowing that has happened. It's a euphoric feeling, it truly is.

For his heroic actions and stalwart leadership during the Battle of Nasiriyah, Gunnery Sergeant LeHew received the Navy Cross, second only to the Medal of Honor.

"The true hero of this family is my wife, Hospitalman First Class Cynthia LeHew," he told me. "She has been an active duty corpsman in the U.S. Navy for nineteen years now and has kept this family together for all of those combat deployments, to include going out herself in 2006–2007 while I stayed home and played a very bad version of Mr. Mom to our daughter, Aisley."

Gunnery Sergeant LeHew mentioned the often-unnoticed toll that years of deployments have taken on military families. "The sacrifices that military families make in the service of this nation is unrivaled, and I have many, many friends whose marriages, for one reason or another— mostly due to multiple, long periods of separation—did not last," he said. "It takes a very strong and independent individual to stand the test of time in a successful military marriage."

Years later, the Marine who once led young men in some of the harshest combat in decades spends his weekends playing with his grandson, Christopher, and counting his blessings.

"With all that we have survived, we take absolutely nothing for granted and consider each day a gift," he said. "Friends come and go and we make many acquaintances in our lives but family ... family is forever."

Sergeant Major Justin LeHew (*U.S. Marine Corps*)

CHAPTER THREE

"IT WAS LIKE A
HORROR MOVIE."

ERIC GERESSY
ARMY COMMENDATION MEDAL WITH VALOR, *BAGHDAD*

*A*merican, British, and other coalition forces rapidly made their way across southern Iraq in late March and early April of 2003. Along the way they met a patchwork of fierce resistance, surrendering and retreating Iraqi soldiers, and jubilant civilians, mostly Shiites, who had been oppressed by Saddam's regime for decades. Many greeted our warfighters as liberators, and often cooperated with their efforts. But cooperation sometimes brought severe retribution by Saddam loyalists. "Children as young as four have been snatched from their parents and hung from lampposts or burned alive ... as a way of punishing their parents," read a report in the Daily Mirror. The article also described how in the southern Iraqi town of Basra, a father who had shared a laugh with British soldiers was forced to watch as extremists burned his son alive.*

Saddam's reign of terror, however, was coming to an end. After nearly three weeks of heavy fighting and lightning-fast advances, U.S. and coalition forces reached Baghdad in the first days of April 2003. As families were gathering for church services in America on Sunday, April 6, their local newspapers carried a story from the Associated Press that described how our warfighters had "rolled through the streets of Baghdad in armored vehicles yesterday, as missiles screamed through the skies and the crackle of heavy machine gun fire grew more intense."

Saddam's forces crumbled and ran away. It was only a matter of time before Baghdad fell, and everyone knew it except the insulated regime that had held such a firm grip on Iraq for decades. They were in complete denial.

"Be sure that Baghdad is safe and secure and our great people are strong," said Iraqi information minister Mohammad Saeed al-Sahhaf, in an April 7 news conference. He was nicknamed "Baghdad Bob" and made infamous by his ridiculous statements. He told reporters that American troops were beginning to "commit suicide on the walls of Baghdad. The soldiers of Saddam Hussein have given them a lesson they will never forget."

Two days later, on April 9, even Baghdad Bob had to admit defeat, but television cameras couldn't find the once confident spokesman. He and other Saddam loyalists had gone into hiding. American forces had taken Baghdad, the Iraqi military took off their uniforms and melted away into the general population, and many residents took to the streets in celebration. One iconic scene from that afternoon occurred in Baghdad's main square, where sledgehammer-wielding citizens attacked a towering statue of a waving Saddam Hussein.

A reporter for the BBC was there, and wrote that the men "scaled the statue to secure a noose around its neck but were unable to pull it down. Then U.S. troops joined in, and used an armored vehicle to gradually pull down the statue." The reporter then described how a soldier covered Saddam's face with an American flag. "The crowd did not welcome the move ... it was replaced by the old Iraqi flag, to roars of approval. As the statue fell to the ground at last, the crowd surged forward and jumped on it. Chanting and jeering, they danced on the fallen effigy, kicking it and hitting it with their shoes in a symbolic gesture of contempt as it was torn to pieces. They then severed the head, tied chains around it, and dragged it through the streets."

A few days later the Washington Post described how "not all the passion was joyful" and how many Iraqis were far from welcoming. "Some Iraqis wept bitterly at the sight of Western troops, not from love of Saddam Hussein but from shame and humiliation."

One of the American warfighters who was moving on Baghdad at that time was Eric Geressy, a thirty-two-year-old sergeant first class in the 101st Airborne Division under the command of then-Major General David Petraeus. Geressy is an old-school soldier from Staten Island, New York, the kind of man who is literally the backbone of our Army ... and he had a score to settle.

ERIC GERESSY: My mother, father, and aunt were all at the World Trade Center on September 11, 2001. My mother, Maryann Rockett Geressy, worked in the World Financial Center, which was across the street from the World Trade Center towers. She was late for work that morning and got there right as the second plane hit. When she came up from the train, there was a piece of the wheel from one of the airplanes in front of her. She had to watch a bunch of those people jump from the towers, too. Then she was in the mad dash trying to get out of New York City. My father, Vincent Geressy, worked for the police department at One Police Plaza. He volunteered to go and do all the cleanup and look for anybody that might have survived.

My aunt, Patricia Geressy, was in the first tower to fall. She worked for the Transit Center, who supported the New York/New Jersey Port Authority in 2 World Trade Center, which was located on the twenty-second floor. While drinking coffee with her coworkers, they felt the building jolt when the first plane struck 1 World Trade Center. She then called the police in the building and asked what had happened, and she told them that they could see all kinds of papers and debris falling outside. The police told her there was an explosion in 1 World Trade Center and to leave the building, so that is what she and her coworkers did. Immediately after she got off the phone, the second plane hit her building. My aunt and her coworkers gathered their things and then walked down the stairs. While moving down the stairwell a voice came over the intercom system saying that 2 World Trade Center was secure and to go back to their desks. Thankfully, they ignored the announcement and proceeded to leave and make their way to the Staten Island Ferry Terminal. Once they arrived at the terminal building, 2 World Trade Center collapsed. She survived, but has never forgotten the events of that day.

So after what happened in New York, I wanted to do my part. I was stationed in Hawaii at the time, with the 2nd Battalion, 27th Infantry Regiment—the Wolfhounds. They are a great unit with a proud history, but after the 9/11 attacks we were tasked with guarding military housing areas on the island. Although that was a very important job, I desperately wanted to get back into one of the airborne units and do my part in the war that was coming. So I spent several months calling the U.S. Army Human Resources Command trying to get back into an airborne unit, and I finally received orders for Fort Campbell, Kentucky. I was a sergeant first class at the time and was assigned as the platoon sergeant for Second Platoon, Bravo Company, Second Battalion, 187th Infantry Regiment, 101st Airborne Division—the Rakkasans.

The 187th was originally a glider paratroop regiment for General Douglas MacArthur that fought in the Pacific theater during World War II. They jumped into the Philippines and were then part of the occupation force in Japan. While in Japan they did a few training jumps, and the Japanese civilians called them "Rakkasans," which loosely translates to "falling down umbrellas," because of the way a parachute looks coming down. The name stuck and it's still part of the regiment's history. We even have an image of a traditional Japanese gate, called a "torii," on the side of our helmet. Once we ran into a bunch of Japanese soldiers in Kuwait and they saw the torii on our helmet. We all yelled, "Rakkasan!" They were shocked. It was kind of funny.

The platoon sergeant is the senior noncommissioned officer for a platoon. We were in an air assault rifle company which had about thirty-four soldiers assigned to each rifle platoon. I was responsible for all of the "skill level one" training. Those are the tasks that enlisted infantry soldiers are required to do under combat conditions. The NCOs [*noncommissioned officers; the corporals, sergeants and above*] are where the rubber meets the road in a unit. The NCOs' main responsibilities are training soldiers for combat and then leading them in combat. That's the job. The other duty you have as a platoon sergeant is to assist in the development of your platoon leader, who is an officer. We were lucky; our unit had some great lieutenants who always put the troops first and always listened to and trusted their NCOs.

We got to Kuwait in January of 2003 and did our final preparations in the desert at Camp New Jersey. We went to an assembly area right on the Kuwait-Iraq border and attacked on the first night of the invasion, ending up at a place called Forward Operating Base Exxon, which was about thirty kilometers south of Nasiriyah. We set up a platoon defensive perimeter, digging in and pulling security for all the helicopters that were there—the Apaches and others. They were conducting deep strike attacks into Iraq.

We were there living in the holes that we dug in the middle of the desert for about three weeks and then finally got orders to conduct an air assault. It was into an assembly area so we could attack the city of Al Hillah, also known as Babylon. After the attack and securing of our unit objectives the next morning, we started the movement north towards Baghdad. We got into another assembly area for the night, and the next morning my company commander and first sergeant came back from a meeting and told us that we would attack Baghdad and would be moving out in the next hour. So we sat down, wrote a quick plan, and then started our movement by truck to our objective, which was one of the air defense artillery batteries on the southern outskirts of Baghdad.

After clearing and securing our platoon objective—the air defense artillery site—that night, my platoon leader and I were checking the perimeter to ensure we had 360-degree security for the night and to prepare for any counterattacks. While we were doing this, our battalion headquarters was also moving in to use the compound for their command post. As we were checking, I suddenly saw a bunch of guys setting up lights. I first thought it was soldiers from our headquarters company. I started yelling, "Hey! Shut the f'ing lights out! What the hell is this?" There was a guy on one knee, facing away from me. He turned around and looked up—it was Geraldo Rivera! I'm like, "You've got to be kidding me. Where the hell did you come from?"

He looks at me and says, "Hey. How you doing?"

"First, can you shut the damned lights out, that's how I'm doing," I said. "Second, what the hell is going on? What's the news?"

Geraldo said, "What's the news? You guys are the news. You just took Baghdad!"

Geraldo stayed with my platoon for about a week after that and took pictures with all of the guys, too. Geraldo is as much of a wild man in real life as you see on camera; he was really good to my soldiers.

Eric Geressy with Geraldo Rivera, outskirts of Baghdad,
April 2003 (*E. Geressy*)

After that we spent about three days doing raids into downtown Baghdad. Our company then moved into the Medical City section of the capital. As we moved by truck into Baghdad, the streets were lined with people cheering and clapping. It was like a scene from that street in Paris after World War II, the Champs-Elysées. People were cheering like crazy for us, but I suspected it would only last for a short period of time.

The city was just beginning to realize it was free of Saddam's regime. An Iraqi named Yasser Alaskary described to a reporter from a British newspaper: "My aunt put out a black banner—an Arab mourning ritual—with the names of all her relatives who had been murdered by the regime ... she looked down her street, and there were black banners on almost every house. On some houses it looks like a long shopping list. Under Saddam it was a crime to mourn people killed by the regime. Everyone was suffering terribly, but they were suffering alone."

Many, even those who worked for the regime, couldn't believe it was ending. The New York Times carried an account of a thirty-nine-year old Iraqi man who had been assigned by the regime to follow foreign reporters around the city. He stopped in front of Iraq's Olympic Committee headquarters, an infamous location that was used to torture athletes who failed to win games and medals for the regime. "Touch me, touch me, tell me that this is real, tell me that the nightmare is really over," the man said, tears running down his face.

Once in the city for good, we initially stayed at the Republican Guard officers' club, and we slept in a parking garage for a couple of nights. That was nice because there was a lot of shooting at night and we had cover. It was still the "wild west" in the city.

Our mission was to secure the hospitals and prevent them from being looted, but there was a lot of sectarian violence happening as well. It was not widely known as "sectarian violence" back then, but that's what it was. People were going after Saddam loyalists, especially at night. You'd hear a lot of gunfire. They were shooting at each other, and they were shooting at us, too.

Medical City was, at that time, host to the premier medical facilities in all of the Middle East. It was a series of five or six hospitals—a surgical hospital, a children's hospital, a teaching hospital, and facilities like that. They were all connected through a large complex of underground tunnels. There were dead bodies all around. People were leaving them outside of the hospital and everywhere else. Next to our platoon's checkpoint there was a truck trailer with a very bad smell coming from it. We opened it up to find a dead body on a litter. The Iraqi doctors told us he was a Syrian who was wounded fighting the Americans, probably in the fighting at Saddam International

Airport. After he died at the hospital, they placed his body in the truck ... in the 120-degree heat for about three weeks. I told the doctors, "Hey, that's enough already. Do something with his body." Some workers from the hospital came out and buried the body in the grassy median on the street. I guess that was better than letting him sit in the truck. It was all just a bad situation ... the doctors were overwhelmed.

It was also our platoon's task to go through those tunnels connecting the hospitals. We had to clear and secure them. I remember one day going through the tunnels ... we found sheets with body parts wrapped inside, arms and legs. There were maggots all over the place, blood, and the smell of diesel. It was like a horror movie.

The doctors were all Shia, and I remember one day some people brought in a pregnant Sunni woman who was all shot up. The doctors didn't want to operate on her because she wasn't a Shia. I spoke to the doctors ... rather harshly ... and convinced them that they were indeed going to treat her. They did, but it was crazy. It wasn't how I envisioned things were going to be, but I guess that's how every war is.

The hospital was hot. There was no air conditioning. The power was barely working, only emergency power, really. There were regular patients, too. Everything that goes on in a regular hospital, during regular times, was still going on there. People still got sick. Women still had babies. Everything was still happening as usual, but there was also a war happening. I remember the doctors were really thankful that we were there. They said if we weren't there, the hospital would have been looted and they wouldn't have had the ability to take care of the people. I also remember how curious they were about us because many of them had never seen Americans before. They thought we were going to come into Baghdad, kill everybody, and destroy the whole place. But when they got to meet us and see us, they thanked us.

There was one doctor in particular who was educated in the United States. He could speak English perfectly and asked us one day, "Could I bring down a couple of the nurses? They have never seen

Americans." So he brought these nurses down and they said hello. I remember they gave me big hugs and thanked us for protecting the hospital. That made me feel good because when we deployed it was all, "Get the weapons of mass destruction," and "Get al-Qaeda," and this and that. I didn't know what to expect, but there we were, spending the war with normal everyday people who were just trying to live. We were there with them, trying to do some good. There was a lot of ugly in the three tours I spent in Iraq, but that time at Medical City is something good I can look back on.

Eric Geressy outside of the Baghdad Teaching Hospital, April 2003 (*E. Geressy*)

Living conditions were absolutely miserable, though. At that point, from the start of the mission in Kuwait to the seizure of Baghdad, we didn't get a shower for something like fifty-eight days. Our uniforms were filthy and we were all very dirty. A lot of people were getting sick, throwing up, and a lot of dysentery was going around. At one point, about three-quarters of the company was throwing up at once. And it wasn't like, "Oh, I'll be taking the day off because I'm sick." I remember throwing up for hours and hours, and then leading a patrol on a raid when it was 120 degrees while wearing heavy body armor and carrying pounds of weapons, ammo, and other equipment.

We set up our company inside the headquarters building for Medical City, in an administration building called Saddam Auditorium. It was a fairly large building, but the bad part was that it was a glass building right on the Tigris River. It was an easy thing to shoot. We had broken up our platoon, about thirty-five or thirty-six soldiers at that point, into two twelve-hour patrols so we could maintain a twenty-four-hour presence in the area. I led a patrol for twelve hours with two squads and my lieutenant took the other twelve hours with the other two squads. I don't know how I got talked into taking the day shift because that time of day was crazy. People were constantly swarming us with questions, asking for permission to do just about anything. They had no concept of being free to do what they wanted, so they were constantly walking up to me and asking, "Could I have permission to do this or that, or could I have permission to leave the country?" I'd say, "You're free. You can do whatever you want to do." They'd say, "Really? So you telling me I can leave?"

I remember one lady came up to me during a patrol and said, "My son was in the Gulf War in 1991. They took him here to Medical City but we never saw him again. Could you help me find my son?" I wanted to do something for her, but she hadn't seen her son since 1991. I knew that there was no way that guy was still alive. But I told her we'd try to help, and that we'd look around and ask the doctors. It was stuff like that all day long. One thing after another. We tried to help, but the conditions were so bad for everybody. There wasn't much we could do.

Across the once shackled nation, newly freed people were emerging from the shadows of fear and oppression to reclaim their liberty, and the loved ones who were imprisoned and tortured during Saddam's reign. They began searching jails, prisons, hospitals, and the dreaded underground chambers used by the Iraqi intelligence service and secret police forces. "These hellish tunnels, with their dungeons and torture chambers, have become infamous, though few ever emerged to give first-hand accounts of life below ground," wrote one Canadian newspaper reporter who saw the tunnels. *"Although U.S. forces have made a concerted effort to find and search this netherworld for survivors, the families of the missing have grown desperate, maddened at the thought that their relatives may have survived incarceration, only to be drowned by flooding pipes or starved to death since their captors fled the city."*

A father spoke with a *New York Times* reporter about his son who was arrested in 1991. *"I want to tell the people of America and Britain something,"* the grieving man said. *"There is nothing, nothing more terrible for a father and mother than to have their child taken from them. Not to know. Never to see his body. You cannot imagine. This is how we lived."*

Author and essayist Christopher Hitchens, a man of the left who endured the scorn of his comrades by supporting the war, wrote years later in his memoir of a particularly gruesome yet revealing afternoon he spent near the ruins of Babylon after the invasion. The locals had descended upon a once forbidden patch of land where in 1991 they witnessed *"three truckloads of people, three times a day, for a month"* being driven there. They were forced to dig their own graves, and were then shot into them or buried alive. Hitchens picks it up from there: *"Seizing the chance to identify their missing loved ones, local people had swarmed to the place as soon as Saddam's regime disintegrated, and uncovered three thousand bodies with their bare hands before calling for help from the coalition."*

Hitchens continued: *"Lines of plastic body bags were laid out on the ground, sometimes 'tagged' with personal items and identifying documents. Where digging was complete, the ground had been consecrated as a resting place. Elsewhere, the ghastly spadework continued. The two men in charge of the scene were a Major Schmidt from New Jersey and Dr. Rafed Fakher Husain, a strikingly composed Iraqi physician. 'We lived without rights,' he told me with a gesture of his hand toward this area of darkness. 'And without ideas.' The second sentence seemed to hang in the noisome air for longer than the first, and to express the desolation more completely. There were sixty-two more such sites, I was to learn, in this province of southern Iraq alone."*

Similar scenes were playing across the nation. Geressy and his soldiers would see Iraqis asking for help routinely during their patrols, but there *was little comfort our warfighters could provide these families, and could offer even less help with actually finding their lost relatives.*

So that's how the day patrol was. At night was when the shooting would happen. Every night like clockwork we'd be sitting inside Saddam Auditorium and we would take incoming rounds from a certain area across the Tigris River. It was AK-47s and machine guns, most likely RPK or PKM machine guns. No RPGs, though. Just small arms, but it was only a matter of time before our guys were going to get wounded or killed. It was late April by that time, and I finally had enough of it. So when I came off patrol one day I asked my company commander, Captain Bret Tecklenburg, for permission to take my two squads and set up an ambush on the auditorium's roof. Then, when we got shot at that night, we could engage the enemy and hopefully put a stop to it before any of our guys got killed or Iraqi civilians were killed.

So I took two squads: one of my rifle squads and then a machine gun squad. The rifle squad had nine soldiers, led by a staff sergeant, and two four-man fire teams with two squad automatic weapons, M249s. They had two grenadiers with M203 grenade launchers. Two riflemen carried an M4 carbine and then the team leaders carried M4 carbines, as well. The machine gun squad at that time had two M240 bravos, which shoot a 7.62-millimeter round, a bigger bullet than what our other weapons fired. They are belt-fed machine guns capable of putting a lot of firepower downrange to suppress the enemy. You use a rifle squad and machine gun squad like this: if the enemy is in a building firing at us with AK-47s or RPGs, we can use the machine gun squad to suppress the enemy, to keep the enemy's head down while we would maneuver and flank them with the rifle squads. The rifle squads would enter the building, maneuver on and then kill the enemy.

On April 26, we went on our normal twelve-hour daytime patrol and came back around 1800 or 1900 that evening. I let everybody eat dinner, and then we transitioned for nighttime operations. That consisted of getting our night vision goggles mounted and ensuring that our laser-sights were in working order. We then went to the roof, got everyone into position, and then waited to get shot at. So, yeah, everybody was pretty happy about that. [*Laughs*] My plan was simple: I set the guys in positions to over-watch the buildings on the other side of the river, which is where we were getting engaged from damn near every night. I had the machine gun squad on one side and the rifle squad on the other. Once the enemy fired at us, we would pinpoint where the fire was coming from, and then engage.

As the night went on, we started seeing enemy movement on the other side of the river. It was about thirty fighters, but they didn't realize that we had laser-sights on our weapons and that we could see in the dark. We also had infrared searchlights that were attached to our weapons, allowing us to light up the whole area, but only we could see it through our night vision goggles. The enemy was moving around in the alleyways setting up positions, and we were watching the whole time. They had absolutely no idea that we could see them. I walked over to speak with the rifle squad leader, Staff Sergeant Scott Heaward, to check their position, and as I was talking with him all hell broke loose. We started getting fired at, a little at first, and then a large amount of fire all at once.

The rules of engagement were pretty tight at that time. We were under orders not to fire across the river, normally, because that sector was supposed to be secured by another American unit. We didn't want to commit fratricide by shooting at the area patrolled by other U.S. forces. So, with that order still on their minds, initially my guys did not return fire.

The fire was coming right over where my machine gun squad was, so I started running across the roof to their position. It felt like the enemy was aiming right for where I was running. I remember thinking that if I would have put my hand up, my arm would have been shot off. I was running across the roof and one of my machine gun squad leaders, Corporal Christopher Galindo, was on the gun and directly across from the enemy. He looked back at me as I was running to his position and screamed, "Hey, what should I do?" I yelled, with some colorful expletives included, "Get the gun in action!

Engage before I get on the gun myself." He turned around and immediately started firing at the enemy position. Within about three bursts Galindo had taken out the enemy machine gun position with great accuracy. We didn't take any more fire from that position.

We mixed it up for another couple of minutes. They were firing from five or six different positions, in buildings and from alleyways. But we cleaned them up real quick. The whole thing was probably about a four-minute engagement, and that was it. From that point onward the hospital never got shot at again. Ironically, a couple of the enemy's wounded fighters were brought to the hospital the next day. They said about fifteen in their group were killed, and many others were wounded, so my machine gunners and my rifle squad did a really good job.

During that time period, we had no way of knowing what was going on with the war outside of what our platoon was doing. Somebody donated satellite radios to the company, privately, and I remember listening to the BBC and CNN International and then letting my guys know what was happening. That's how we knew what was going on in Iraq ... on the radio, and we were actually right there.

The next week after the firefight at the hospital we were having breakfast in the Saddam Auditorium and heard President Bush give his "Mission Accomplished" speech from the deck of the aircraft carrier USS *Abraham Lincoln*.

The speech occurred on Monday, May 1, after President Bush landed on the carrier in a S-3B Viking jet. "Major combat operations in Iraq have ended," the president said. "In the Battle of Iraq, the United States and our allies have prevailed. And now our coalition is engaged in securing and reconstructing that country."

I remember we were sitting there eating, all filthy dirty, having just had a big firefight. I looked at my guys and said, "Apparently the enemy didn't get the memorandum that the war is over."

Top military planners began making preparations for most of our forces to return home during the summer, leaving only, by some accounts, two or three divisions to secure the entire country. Our leaders seemed eager to cash in the "peace dividend" that usually follows the cessation of hostilities. But the war wasn't over; in fact, it had hardly just begun.

Sergeant 1st Class Geressy would remain in Iraq for another ten months, eventually moving with the 101st Airborne Division to northern Iraq and the city of Mosul. There, Major General Petraeus would earn acclaim for bringing security and stability to the area through what would eventually be recognized as classic counterinsurgency methods. Elsewhere in Iraq, however, the situation was growing dim as the military began to prepare for a major withdrawal ahead of the reconstruction phase. But as warfighters like Geressy were seeing on a daily basis, and as the American people would soon learn, there was no widely circulated or practically implemented plan for the post-invasion period known in the military as "Phase IV Operations."

Later that year, the Washington Post quoted a well-respected junior officer as saying "there was no Phase Four plan" for the occupation. "While there may have been 'plans' at the national level, and even within various agencies within the war zone, none of these 'plans' operationalized the problem beyond regime collapse," said Major Isaiah Wilson. Although only a major at the time, Wilson was a credible witness; he spent the first few months of the war as a researcher with the Operation Iraqi Freedom Study Group, poring over plans and orders, and then he spent the early days of the insurgency as the chief war planner on the staff of Major General Petraeus when he commanded the 101st Airborne Division. He concluded that the U.S. military was "perhaps in peril of losing the war, even after supposedly winning it."

Geressy kept fighting, and eventually received the Army Commendation Medal with Valor for his actions on the rooftop in Medical City. The citation read, in part, that the Staten Island native was "fearless in the face of the enemy." He would confront that enemy several more times before the war was over, and earn more accolades for his valor.

Writing only months after the invasion in his book, The Iraq War, British military historian John Keegan noted "the war was not only successful but peremptorily short, lasting only twenty-one days, from 20 March to 9 April. Campaigns so brief are rare, a lightening campaign so complete in its results almost unprecedented ... The Iraqis had fielded a sizable army and had fought, after a fashion. Their resistance had simply been without discernible effect. The Americans came, saw, conquered. How?"

The premise of Keegan's question was premature and the answer is clear in hindsight: The fight wasn't finished at all, and what became known as "the insurgency" was about to begin.

PART II

THE INSURGENCY, 2003–2006

CHAPTER FOUR

"SHUT UP AND SHOOT!"

JEREMIAH OLSEN

SILVER STAR, *BAGHDAD*

I raq was in crisis. *The regime had fallen, its military and police forces had abandoned their posts, the civilian bureaucracy had vanished, and nearly every public service and institution, business, market, gas station, and every other entity required for normal daily life had collapsed into the vacuum. All that remained was a mixture of confusion, hope, fear, celebration ... and retribution. Our military had smashed Saddam's forces, but now it was our responsibility to keep Iraq from tearing itself apart amid lawlessness, power grabs, and revenge killings.*

In early May 2003, just a few days after the president's "Mission Accomplished" speech, the White House issued a statement that a little-known former State Department diplomat had been appointed as our top official in Iraq. Ambassador L. Paul Bremer would serve as the "Presidential Envoy" and "oversee Coalition reconstruction efforts and the process by which the Iraqi people build the institutions and governing structures that will guide their future." The organization that the ambassador would oversee, the Coalition Provisional Authority, was responsible for everything— governance, the economy, utilities, transportation, education, housing, medical care, and anything else that a government either provides or regulates. Everything, that is, except security. Military operations would remain in the hands of the American generals.

In the early days of the war, our invasion and occupation of Iraq was often compared to our experiences in Germany and Japan after World War II. We wanted similar results, but nothing about the wars was similar. Those nations were defeated in 1945, their armies surrendered, and their people unconditionally submitted to foreign occupation for several years. We didn't fight like we did in World War II, nor did we have the same objectives, so we shouldn't have expected similar outcomes. In Iraq, we only toppled a leader and his cronies, not a nation. The army never actually surrendered and the people weren't willing to take advice, much less orders, from people they considered outsiders at best, or infidel invaders at worst.

Every action by the coalition was met with growing resistance. One of Bremer's earliest and most criticized orders was to disband the Iraqi military, intelligence services, and militias, and to purge senior Baath Party officials from public employment.

Some of our military leaders were shocked by the order. "None of this was coordinated with the U.S. military," wrote retired Lieutenant General Daniel P. Bolger in his book about the war. That's partly debatable, as many accounts have shown Bremer's order was circulated at the Pentagon, although likely not as widely as necessary for everyone to voice their concerns about the document. The commander in charge of the invasion, General Tommy Franks, would have surely opposed the order because his plans had always relied upon the Iraqi army to keep order.

Regardless, there wasn't an existing military to "keep order," as Franks may have hoped. Divisions of the Iraqi army didn't line up in ranks to watch their commanders surrender to the Americans, and then await orders for their next mission. They abandoned their bases, took off their uniforms, and melted into society. Their installations were ransacked and their equipment was looted. The Shia and Kurds would have strongly opposed bringing the Iraqi military back. Still, Franks and others hoped to recall the army—or whatever remained of it—and use it to restore order and augment the shrinking size of the U.S. military footprint. There were rumors that portions of the army were ready to come back, but Bremer's order officially took that option off of the table.

The process to weed out top Baathists from power was equally criticized because it removed too many potential leaders from the bureaucracy and probably pushed them into the coming insurgency. American political leadership saw it as a necessary act, however, if Iraq was ever to emerge from Saddam's shadow. His Baath party had been in power much longer than the Nazi party held power in Germany, and its tentacles had reached deeply into every sector of society.

"Baath Party members were required to attend weekly indoctrination meetings where they had to memorize the latest party slogans eulogizing Saddam," wrote Bremer. "Members were expected to recruit children ... party members were required to spy on their family, friends, neighbors— and fellow Baathists. These dehumanizing practices, combined with Saddam's and his sons' capricious brutality, had created an atmosphere of pervasive fear and mistrust throughout Iraqi society."

Most everyone agreed that those who had committed serious crimes during Saddam's reign of terror shouldn't have been allowed to return to government or military service, but the order cut far deeper than the senior levels. Rather than judging Baath party members individually, the order banned all those who had achieved a certain level of membership. Included in the ban were people like university professors and mid-level bureaucrats, who claimed to have only joined the Baath party out of economic necessity. Bremer noted that the "administrative inconvenience" caused by banning these individuals from public life would eventually be eased because, in his estimation, "apolitical technocrats were usually the people who made organizations work."

While purging the most rotten Baathists from power may have been necessary, the "apolitical technocrats" failed to emerge amid the growing insecurity. Many simply feared being killed if they worked with the Americans, or even went back to their old jobs in the Iraqi government. We couldn't have expected a foreman to return to his job at the utility company if doing so would cause his family to be murdered.

So the nation remained in a broken state of crisis, with these two mis-calculations—the order to disband the military and to purge the government of senior Baathists—becoming emblematic of the next few years in Iraq. Bolger later wrote that "de-Baathification was announced as a fait accompli. It guaranteed Sunni outrage."

Ultimately, Iraq was an enraged, broken society long before the invasion, and it was only held together for so long by Saddam's brutal grip. Now that his regime was gone, Iraq was less of an actual nation than a collection of groups with justifiable grievances against one another, a deep and paranoid suspicion of everyone's motives, and an understandable reluctance to surrender their newly obtained freedom by trusting the Americans or another faction with power.

Former Baathists weren't driven from power everywhere, however. The only large city in Iraq that was seeing any level of relative peace at that time was the northern city of Mosul. There, Major General David Petraeus was implementing a counterinsurgency-like strategy, and had managed to secretly have his area exempted from the de-Baathification order so he could freely select Iraqis for positions based upon competence, diversity, and balance. Other exemption requests were denied, and the order cut wide and deep across the country.

Bremer and his Coalition Provisional Authority would spend the next year in a desperate effort to keep those factions from pulling Iraq into deeper chaos. The orders to disband the military and purge senior Baathists kept the Shia and Kurds aboard, but eventually significant portions of the Shia faction would wage war against the Sunni insurgents and the coalition for years to come. Some believed, as time would eventually prove, that the war simply wasn't over yet. Scores remained to be settled. Winners and losers in the conflict had yet to be decided. And until those who wanted or were willing to fight had been killed, exhausted, or clearly defeated, the insurgency would rage on.

Meanwhile, our military was still very much in the fight. Aside from trying to establish some law and order while beginning to train a new Iraqi military and police force to replace them, coalition forces were actively seeking two objectives whose acronyms would become synonymous with the era: HVTs, or "high value targets," and, of course, the WMDs that had partly lured us into the war to begin with.

Since the invasion, a special group had been scouring all of Iraq looking for WMDs, but in early May its hopes began to fade. "Leaders of Task Force 75's diverse staff—biologists, chemists, arms treaty enforcers, nuclear operators, computer and document experts, and special forces troops— arrived with high hopes of early success," wrote Baron Gellman in an early

May 2003 edition of the Washington Post. But the group was packing it up and heading home.

The 75th Exploitation Task Force, which was its formal name, didn't find the WMD stockpiles that Western intelligence agencies had predicted were there. Officials blamed many factors, but according to Gellman's article, "the greatest impediment to the weapons hunt ... was widespread looting of Iraq's government and industrial facilities. At nearly every top-tier 'sensitive site' ... intruders had sacked and burned the evidence." The task force's mission would soon be handed over to an entity called the Iraq Survey Group, whose work would stretch into the coming months.

Elsewhere, special operations forces were trying to capture or kill HVTs. While there were rumored to be hundreds of such targets, dozens would become infamous for their inclusion on what became known as the "Most Wanted" deck of cards that were distributed to coalition forces by the U.S. military.

"Cards showing 'ace' of spades Saddam Hussein and 54 other regime members have become hits around the world, especially in the United States," read a report from the Armed Forces Press Service. "So far, 17 of the Iraqi most wanted have surrendered or been taken into custody." The cards became so popular that they were selling in stores in the United States, and the media would report HVTs being captured by referring to their specific card within the deck.

Americans were beginning to learn a great deal about one of the face cards: the Ace of Hearts—Uday Hussein. The eldest son of Saddam was well known inside Iraq for his brutality, which far surpassed that of his father. He tortured athletes on Iraq's national soccer team who failed to perform to his expectations, raped girls and women who were unfortunate enough to catch his eye, and killed without fear of punishment. While many believed Uday's psychopathic behavior excluded him from succeeding Saddam, some feared he would indeed eventually assume the presidency through the force of his father's grip on the Baath Party and Uday's command of the fanatical Fedayeen Saddam, which was a paramilitary band of fighters who swore allegiance to Saddam personally.

In late May, writers Brian Bennett and Michael Weisskopf published a striking article in Time magazine titled "The Sum of Two Evils" that *illustrated the need to capture, or preferably kill, Saddam's sons. It was the first of many articles introducing Uday and his younger brother, Qusay, to the American people. Their story recounted the tale of one of Uday's former aides who said that he arranged a party in 1998 at an equestrian club for his boss.*

During the party, Uday stood atop a building and used binoculars to search the crowd for a target. "Uday tightened the focus on a pretty 14-year-old girl in a bright yellow dress sitting with her father, a former provincial governor, her mother and her younger brother and sister," the article explained. After failing to entice the girl away from her family, Uday's bodyguards simply "picked her up and carried her to the backseat of Uday's car, covering her mouth to muffle her screams."

The girl was returned home three days later, with "a new dress, a new watch, and a large sum of cash." Her father protested publically for weeks, demanding punishment for Uday. The serial rapist and heir-apparent finally sent his bodyguards to speak with the father. They demanded that he not only send the girl back to Uday, but that she must bring her twelve-year-old little sister, as well. After being threatened with death, their father submitted and sent both of his daughters with Uday's henchmen.

Uday Hussein was a monster who needed to be killed, and quickly. Thankfully, he and others on the infamous deck of cards were now being hunted by some of our nation's most highly skilled warfighters, including twenty-four-year-old U.S. Army Corporal Jeremiah Olsen.

Olsen was a member of the 75th Ranger Regiment, whose recruiting website describes its members as "more than just physically strong, Rangers are smart, tough, courageous, and disciplined. Rangers are self-starters, adventurers, and hard chargers. They internalize the mentality of a 'more elite Soldier,' as the Ranger Creed states and as their intense mission requirements demand." In early June, Olsen's heroic actions would prove his regiment's motto: "Rangers Lead the Way!"

JEREMIAH OLSEN: I was born on Whidbey Island. That's north of Seattle, Washington. There's a naval air station there where EA-6B Prowlers and A-6 Intruders were stationed. There is also a lot of history on the islands north of Seattle, making it a nice place to be a kid.

Why did you want to become a Ranger?

It was around the year 2000 and I was living in Seattle, working and skirting the line between trouble and half-heartedly going to college. I figured that I needed to change the way I was living my life. So I thought that if I was going to join the military, I was going to join on my terms, which was to be something within the special operations community. I just wouldn't have been happy doing anything else.

I went to the recruiter's office and said I wanted to join the U.S. Army's Special Forces and become a Green Beret, but there were no openings into that program. So the recruiter said, "Hey, you can go to a Ranger battalion." He showed me a video about the Rangers, and explained their role in special operations. I thought, "Why not?" and decided to sign up.

Most folks get confused between Ranger School and the Ranger battalions. They're two separate things. The Ranger School is just that—a school. Its graduates earn the right to wear the "Ranger" tab on their uniforms. "The" Rangers, however, are a full-time unit—not a school—and they're part of the military's special operations forces. Rangers go through different training, and it's pretty much non-stop throughout their time in a unit, and most go to the Ranger School, as well. Ranger battalion soldiers wear a tan beret that sets them apart in dress uniform.

In the end, being a Ranger was a better fit for me because they are a special operations force that performs mostly direct action missions. That's where I like to be. So, when I was about twenty-two years old, I enlisted. I first went to basic training at Fort Benning, Georgia, then to advanced infantry training, also at Fort Benning, and then I went down the street to airborne school. Once those preliminary steps were completed, I was then dropped off at the Ranger indoctrination course for the selection process. Back then it was a four-week course called the Ranger Indoctrination Program. Now it's called the Ranger Assessment and Selection Program and it's eight weeks long due to more instruction.

How tough was the program?

Most soldiers begin the program at the peak of their physical and mental conditioning. Almost all have just gone through basic, infantry, and airborne schools, like I had. Others had already been soldiers in the regular Army, so they knew the program's reputation for intensity and they trained hard before arriving. Still, we started with 350 soldiers, and after all was said and done, only about forty of us graduated.

I thought that the initial program was going to be the toughest part of being a Ranger. I was wrong. Actually *being* a Ranger is much harder, by far, than the selection process. That's why they try to weed out anyone who isn't mentally and physically prepared for the rigors of being part of a Ranger battalion.

The selection process was intense. A cadre of active Rangers closely watched us at every step of the way. Just like any special operations selection course, it is meant to test the heart and physical ability of the candidates while testing their decision-making process under duress. The cadre set out tasks for us to complete every day, and all tasks had to be accomplished successfully and within the allotted time, or else you're out. It was always, "You must pass this," or, "You must pass that." Everything was pass the individual task or team task, or fail the overall selection. Which is to be expected, but not everybody shows up with the right mindset, and they allow these stresses to break them.

For instance, one night we had to do a pretty long road march in full gear, maybe eight miles, and then start a land-navigation course before dawn. The course itself was long and had very large sections of land between the points we had to navigate from and to, during the day and into the night. We were all tired, hungry, and thirsty, which is how we always felt during the course, I think. A lot of the guys got dehydrated. That caused them to become confused, and then lost, or it caused them to just flat give up. It's a weird thing when you look down and see someone who seemed so ready for selection just give up on life and lay there like they've accepted defeat. But, naturally, everybody has a breaking point, and just like that, they were out of the program.

Some tasks were small and short, while others were large and long. But it was everything in between the tasks that got to people the most. We'd get ready to perform a mission, form up in ranks, and then the cadre would notice that a soldier had something untucked or that some other small thing wasn't correct. "Everybody get down!" We'd hear that and start doing push-ups. And it wasn't just doing a certain number—we had to do push-ups until someone quit the program, or if we were on a hot blacktop, we would do exercises until there was a fresh puddle of sweat under each man. If the puddle evaporated, we kept going. It's kind of comical, actually. I always tried to see the humor, and feed off of the big picture around me. People got too caught up in the pain to see the humor in the actual activity we were doing. It is a game, a serious game, but in the end you never let the game beat you if at all possible. At least that was my weird outlook on the whole thing.

It was always like that: the extreme version of absolutely everything. Once they had us going all day and we only ate a single Meal-Ready-to-Eat, or MRE, early in the morning, which wasn't out of the normal. We were burning through calories and by mid-afternoon we were all very hungry. We were in a field and they were teaching us how to make a modified stretcher out of sticks to carry an injured person off the battlefield. The cadre said, "Everyone, go to the wood line and get a stick!" Everybody sprinted to the edge of the woods, which was at least 200–300 meters away, and did their best to find what they thought would be a good stick. The last guy back would inevitably cause us to repeat the event, of course. So then the cadre said, "Oh, none of those sticks are good enough. Go get more." We did stuff like that all afternoon, until we were drenched with sweat.

It can get pretty cold in Georgia on a late winter night, especially if you're wet from head to toe from sweat, and exhausted. That's when the cadre made a bonfire with our sticks. We couldn't share its warmth, of course. They told us to get out of the firelight and into the wood line. So we sat there, tired, hungry, thirsty, and freezing, all the while watching the cadre enjoy the bonfire from a distance. "This fire sure is warm!" they said. "If anyone is cold, come on out and enjoy the fire!" If you did, you were out of the program. Still, some guys walked from the tree line and up to the fire, shivering and defeated. But on the bright side, they did warm up.

As it got later in the evening, our empty stomachs were growling and the hunger was intense. We were already hungry from the previous days, and this day was worse. That's when the cadre decided it be a great idea to roast hot dogs over the bonfire. The cadre shouted, "Hey! Who wants a hot dog? They're so good. Come and get it!" The smell blew over to the wood line and drove some of the guys crazy. I saw several of my friends who were in better shape than me, who were stronger than me, break at that point. All night long I saw guys walk over, sit down, eat a warm hot dog by the bonfire ... and quit. They packed their stuff a few hours later and were sent somewhere else. Probably Korea. Thankfully I don't like hot dogs, but either way I wasn't falling for any of that garbage.

The lack of sleep was also part of the process of finding out who was mentally tough enough, I think. The Army has a rule dictating how many hours that a school must allow a soldier to sleep. I think it's about three hours a night, but very rarely do I remember sleeping for three hours. Most of the time I only got about an hour's worth of sleep each night.

In the end, the weeding-out process is more about the mind than the body. Our flesh and blood is stronger than we think, and it can take a great deal of pain before it actually stops working. But our minds? Well, they'll break way before our bodies will, and when our minds give in, for whatever reason, that's when our bodies will break, too. For instance, even though your ankle might be medically fine, if your will to fight has gone, you might actually feel like your ankle is broken, like you cannot walk. Your mind will make every excuse in the world to stop.

After that we had to do a land navigation task, working with a map and compass to get from one point to another in the dense woods. I remember making good time and being about halfway through the distance when my body started to cramp up because I was so dehydrated from those brutal push-ups and the stick hunting. It was so bad that I couldn't even bend one of my legs. So I started to drag my leg through the woods, but it kept getting caught on vines and underbrush. It was like trying to drag a pole through thick bushes. Finally, my abs started locking up, and then my sides. I was out of water, and there was nothing to drink but whatever was in the disgusting swamp. I remember thinking, "I can't move." So I sat on a log and tried to gather my thoughts and wait out the cramps.

One of my friends just happened to walk by and he asked, "Hey, are you all right?" I mumbled something about not being able to move. He stood me up and gave me some of his water. "Here, drink this," he said. I took a big gulp and gave it back, still dazed. Then he said, "What are you doing, man? You need to hurry up!"

Just having him stand me up, give me a little water, and tell me to get going had an effect. I snapped out of it. I was like, "Oh, yeah. What am I doing? I just wasted ten minutes sitting here!" My leg was still messed up, but I kept going and passed the navigation task. I owe that guy a beer, to say the least.

So, after four weeks of days like that, and seeing my class shrink from about 350 to about forty or less, I was finally selected to be a Ranger. The 75th Ranger Regiment had four battalions at the time: 1st, 2nd, 3rd, and HHC [*headquarters and headquarters company*]. I volunteered for the 2nd Ranger Battalion stationed at Joint Base Lewis–McChord near my home in Seattle.

The battalions put all of the new guys on a one-year probationary status to see if you've got what it takes, or if you slipped through the cracks to get there, even after all of that weeding-out during the selection program. Which I learned is a great thing, and like in all SOCOM [*Special Operations Command*] units, it's a necessity.

Rangers have a multi-year training cycle and are always preparing—airborne operations, cold-weather operations, helicopter operations, or whatever else. When you arrive at your battalion, you jump right into whatever phase of training they happen to be in. Right off the bat. It's pretty intense. I got assigned to a squad and, as a new guy, I got hassled. Some guys can't take it and they get out. The Rangers, like all SOCOM units, is an all-volunteer unit, and soldiers can leave at any point. It's not a big deal, either. It's hard, and if you can't take it you really shouldn't be there. The Army will just send you somewhere else.

I arrived at my battalion when they were planning to do airborne operations. We'd get on a plane somewhere, fly for a while in air rig parachutes, and then jump into God knows where. I'd land in the woods somewhere and be like, "Where the hell am I?" Sometimes it'd be Alaska, sometimes New Mexico, and sometimes Oregon. It was always fun to try and figure it out before someone leaked where we actually were.

One day in late August of 2001, we got all of our gear, parachutes, rucksacks, and everything else, and loaded onto a C-17 military transport aircraft. We flew from Washington State to Europe, and aside from stopping on the East Coast for fuel, our feet didn't touch the ground until we jumped from the airplane somewhere over Germany.

We all landed in a farmer's field, formed up, and began a long road march to a training area that was about seventy kilometers away. We cut through fields, walked down country roads, through villages, and even walked alongside the Autobahn for a ways. I remember walking through towns and guys would come out of bars with a beer in hand. I was so jealous; with an M-249 necklace [*linked ammunition*] and a heavy pack on my back, I just wanted a taste of that German beer! A car hit one of our guys, too. The German didn't see him at night, but luckily the car hit his rucksack and sent him flying over the roof. He ended up okay; kind of funny, but luckily he was all right.

We finally reached our objective later that next evening, which was a training area for allied forces. We set up perimeter security and prepared for our simulated fight against what's known as an opposition force, or OPFOR. We fought them for a few hours, shooting each other with simulated rounds. We left that area and trained in a few other places, eventually reaching a large training ground near a town called Hohenfels, Germany. That place has notoriously horrible weather, and the conditions are unusually harsh. The locals told us that's why Hitler sent his SS troops to train there; if they could fight there, Hitler figured, then they could fight anywhere.

That's where I was on September 11, 2001. We had been in the woods for several days and everyone was exhausted, dirty, and cold. We were inside cleaning our weapons and preparing for more training later that afternoon when a military intelligence officer came in and said, "Hey, you guys need to watch this." He turned on the television and that's when we saw that the Twin Towers had fallen. We were all like, "When do we leave to go fight? Are we going to war now, straight to Afghanistan, or are we heading back to Fort Lewis?" We were all ready to go, ready to fight. We ended up going back to the States first, though, because we didn't bring everything we'd need to fight in Afghanistan. Or maybe some other command decision was made.

It was a frustrating time for me; I was due for a promotion shortly after that training deployment and scheduled to attend Ranger School. They told me that I wouldn't deploy to Afghanistan and would stay on schedule, attending the next class at the school. I was really angry. I said, "This is so stupid. I'm going to fail out of Ranger School in the first week so I can deploy with you guys." They told me that I'd get kicked out of the battalion for that. It was a catch-22.

So, the battalion deployed to Afghanistan, and I went to school in Georgia. It sucked. They didn't get into too many tussles, though, so I didn't feel so bad. Now, if they had gotten into a lot of big firefights without me, then sure, I would have been pretty angry to not have been there for my guys. The battalion returned later that next year. We did some more training and then the Iraq War was getting ready to kick off. So we packed up and shipped out.

The 2nd Ranger Battalion deployed to the region around Iraq and staged in an area just across the border. From there we did pre-invasion incursions, going in at night and then coming back. As the invasion drew closer our incursions became bolder. We flew in C-130s into the country, but at very low altitudes. Those trips were similar to carnival rides, and we placed bets on who would puke first.

When the war kicked off, our task force's mission was to search for WMD sites during the day and then HVTs at night. The deck of cards came out with a bunch of HVTs on the faces. Saddam was the Ace of Spades, I think. We went after all of them, and some other guys, too.

I was a team leader for machine gun squad. For us, being able to do what we had trained so hard for and help out in the big picture was a great thing.

Could you describe a typical mission?

Well, one night, after the deck of cards came out, we went on an HVT mission. My team flew out from our base, landed on the target's house, and quickly grabbed a couple of guys before they knew what was happening. As I secured the back of the house I noticed a guy running off into a field of palm trees. I was like, "You're kidding me, right?" So I chased the guy down, tackled him, and then flex-cuffed him. I checked him for weapons, and then dragged him back to the house. He was acting important and running his mouth so I told him to shut up and then placed him against a wall. A few minutes later a helicopter came for the prisoners, then we flew back to base.

The helicopter landed back at the airport and my guys and I were walking back to the hanger. Then my platoon sergeant walked up and said, "Hey, Olsen, some guys want to talk with you." I thought, "I'm in trouble for smacking that guy I grabbed." We went to a little building on the other side of the airport. I didn't know who they worked for but they brought me into a room and one of them said, "We watched your mission from satellite video, and saw you chase that guy down and catch him. It's good you didn't let him get away. He's a very bad guy." Then they told me who he was. Turned out that he was pretty high up on the deck of cards, and they assured me that I could have broken his arm or leg for good measure; he at least deserved that.

That was crazy. You never know who you're going to snatch sometimes. It could be nobody, or it could be somebody. When you're doing a mission like that, in the moment, it's just your job. You're using the skills you've acquired to make sound decisions to accomplish your objective and make sure your guys come back safe. Now, looking back, I would never go about military service in any other way. I don't know how I'd feel about my service if, after all of that extreme training, I never got to put any of it to use. Flying out on helicopters, sliding down ropes and experiencing the wars in Iraq and Afghanistan like I have … I am just grateful that I got to do it, and that none of my guys got killed while I was there.

Tell me about June 11, 2003.

We did a raid targeting a large number of individuals on the night of June 11, far out in the wilderness area, and nowhere near any cities or towns. It ended up being a pretty big firefight that lasted into the early morning hours of June 12. One of my friends lost half his leg, and a lot of enemy fighters were killed, too. That specific part of the operation may still be classified, but what happened the following morning isn't.

After the raid, we secured the detainees and waited until sunrise. That's when the larger, regular army forces would arrive to pull security on the area and perform what's known as site exploitation [*looking for intelligence and evidence*]. So until then, a few other guys and I pulled away to the perimeter to provide security. Then we waited.

I was the truck commander of a Humvee. Our vehicles were set up very differently than trucks are now. They didn't have any armor; they were stripped of everything but the windshield and the roll bar, on which we mounted either a .50-caliber machine gun or a MK19 belt-fed 40mm grenade launcher. We had two or three machine gun mounts on swivel arms on the back and sides, and then we had our own personal weapons. There was no shielding from bullets and there was no shielding from being seen by the enemy; there was nothing. But, the beauty of that configuration was that the Humvee moved faster—if it's possible for a Humvee to move fast—and anybody could shoot at any time, and from anywhere on the vehicle.

The sun rose over the desert and I stood up in the turret of the truck, manning the .50-caliber and looking for anything unusual. It was early June so it was warm but hadn't yet reached the sweltering temperatures that it'd rise to later in the day. It was dry and kind of nice, actually.

The first contingent of regular Army guys started arriving. They were from one of the airborne divisions, but I cannot remember which. I saw them set up a sniper team about 150 meters from my position, so I knew it wouldn't be too much longer before we could pull back. I was looking forward to that because, as usual, we had been up all night. The other guys around me were sleeping, and I was scanning the horizon.

Earlier that morning, Apache attack helicopters were sent out to secure our perimeter and provide over-watch on our objective. The Apache attack helicopters were still flying patrols on the outside of our perimeter to ensure nobody snuck up on us. So I was standing there, in the back of the truck, leaning on my gun and looking out across the desert when I noticed that one of the Apaches had stopped. It was hovering over an area about two hundred meters from my truck. There was a berm between us, sort of rocky ground, so I couldn't see what had caught its pilot's attention. Then a second Apache joined the first, both oriented in the same direction. I leaned in, and remembered thinking, "What are they looking at?"

Just then one of the pilots turned his head. I could see them that clearly—and the gun affixed to the bottom of the Apache whipped over in the same direction, and then let loose a burst of gunfire. I said, "Whoa! What are they shooting at?" I thought for a second that maybe they were just testing their guns. Sometimes they'll do that in the open desert. But then they started shooting their rockets, and I immediately knew it was for real.

Around me at the time was the driver—one of my best buddies who is still in special operations—and about three guys from an anti-armor team with a Carl Gustav [*84mm man-portable reusable anti-tank recoilless rifle*]. I yelled at everyone to get in the truck. The guys from the other team were like, "What? What's happening? Show me where…" but I cut them off. I said, "Just get in the truck or I'm leaving you."

Just as the driver started the truck I saw a smoke trail go toward one of the Apaches and hit it right in the tail rotor. It was a rocket-propelled grenade. The helicopter went straight down, and the other Apache took off, probably because it didn't want to get shot down either.

I said, "Holy shit! We gotta go!"

The guys piled into the truck and we moved in the direction of where that RPG came from. We were hauling ass and saw the crash site as we came close to the berm. We were still about 150 meters away but I could see the helicopter on the ground, burning. The tail rotor was blown off but it was still largely intact, which made me think the pilots could still be alive. I also could see enemy fighters moving towards the helicopter.

My driver started heading closer to the crash site, but a sergeant on the team that was with us, the guys with the Carl Gustav, asked us to stop. He outranked me, but I was hesitant to stop. He wanted to get into a better position to fire the Gustav. So I told him, "I need you to get out of the truck, because I am going to get those pilots, now!" So they got out, and we left them. They were out of AK range, though, so they were fine and safe. The driver then headed straight for the downed helicopter and didn't let off the gas until we got there.

As we pulled up, the driver slammed on the brakes about twenty meters from the helicopter because the fire was causing some of the rounds and rockets to cook off. Just as the pilots were climbing out of their cockpits I saw two, maybe three enemy fighters with AK-47s shooting at us from the cover of a nearby berm. They were about thirty meters from me and about twenty meters from the pilots. It was all pretty close. They were just sticking their guns over the top and firing wildly. I laid down some covering fire from my .50-cal, giving the pilots enough time to run over to our truck. They were both fine, though one was limping from getting banged up in the crash.

I'll never forget what one of those pilots said, a lieutenant colonel, I believe. I can't really remember his rank. He looked at me and, with a 9mm pistol in his hand, said, "Come on! Let's go get those sons of bitches!" I had to smile and kind of shake my head. I said, "All you've got is a pistol, sir. Get in the truck. I'm getting you out of here." The other pilot, the one limping, had already climbed in the truck; the lieutenant colonel was adamant, though. He said, "No, I want to fight!" I said again, "No, sir. I'm getting you to safety. Get in the truck!" He finally got in. The driver threw it in reverse and backed out of there while I kept firing at the guys on the berm.

We drove to the location where I remembered that the regular Army had set up a sniper team. I turned to the pilots and said, "Stay with the snipers. Don't move, and don't get shot; they are fifty to eighty meters in that direction … now go!" They were okay, but I never saw those pilots again. I wonder what became of that lieutenant colonel. I'll never forget that guy. He really cracked me up. I wish that I'd had an extra M4 that morning to let him borrow, but regardless, I wasn't letting him get hurt. It was my job to protect them and get them to safety. I liked his attitude and fighting spirit, though.

Then I looked back to the berm and saw one of the insurgents peeking over the top. I told my driver, "Head straight for that berm where that guy is." We hauled ass, right back at them, but they quickly hid back behind a berm. As we neared the crash site, we saw that the entire helicopter was on fire. Its .25mm cannon ammunition was burning off, rockets were blowing up, and there were hellfire missiles that were lighting off. It was making a whole lot of noise. The enemy had taken cover back behind the berm again, and I don't think they could see us or hear us. I think they lost track of where my truck had gone because of the berm and the noise.

That was our chance to seize the initiative. I decided that we'd drive over the top of the berm and then open up on those guys. Once we crossed the top of that berm, though, we'd be wide open and without armor or backup. I couldn't afford to waste time changing belts or clearing a jammed machine gun. And while we had our personal M4s and sidearms, they weren't good for seriously stopping someone. So I took advantage of the lull to re-link my .50-cal ammunition belts in our large ammunition canisters, or ammo cans.

Our cans were larger than most, thankfully. Before that day we had gone to a few pilots and borrowed the cans meant for their helicopter guns, which are huge cans, much larger than what would normally be found on a truck-mounted gun. I also bartered for their selection of different kinds of .50-cal ammo, which was a better selection than we had. The cans could hold several belts. I then took an extra fifteen seconds or so to double check the .50-cal to make sure it wouldn't jam. I figured either rush into it and get a jam from bouncing in the desert, or take a second and straighten it all out so that there are no problems while engaging the enemy.

When everything was ready, I told my buddy, "Creep forward, slowly, so that I can take careful aim at whatever we see on the other side." He did, and the trucked rolled slowly over the top of the berm. Just as we started to roll down the other side I looked to my right and saw an insurgent standing about ten meters away. He had an AK-47 in his hand and was aiming at the top of the berm. He was caught totally by surprise, probably not thinking we'd come rolling over the berm like that. He glanced up with a surprised look on his face. I swung the turret around and opened the .50-cal up at him. At that range, the .50-cal tore him to pieces.

Then I saw another guy on our right, about fifteen meters from the first. He started shooting his AK so I shot him, too. Another enemy fighter farther to the left started shooting, so I walked the rounds up to him and across his hip. While I was shooting the guys on the left and right, I didn't see the two insurgents who had taken positions directly in front of our truck, behind where the ground gave way to a wadi [*a gully, dry except during rainy periods*] about 8–10 feet deep. They were standing on a rock, holding their guns over the top and firing at us from pretty good cover.

My buddy took out his M4 and started shooting in the same direction that I was shooting. Man, he was one brave dude that day— riding into the fight without a second thought, driving and laying down fire at the same time. He was awesome, but then he got shot. The bullet from the AK passed through his M4's receiver and nearly took off his thumb.

He started moaning, "Ohh! Ohh!" and making a whole bunch of other noises. Again, the guy is one of my best buddies, but we're pretty tough on each other in the Ranger Battalion. I kind of kicked him in the head a little and said, "What the hell are you yelling about? Shut up and shoot!" I was just joking, of course. He looked up at me and said, "Hey, asshole. I'm shot!" I looked down and saw that he had blood spurting all over himself and the truck. I said, "Oh, shit. I'm sorry, dude! Okay, okay! Just put it in reverse and go."

He had started to go into shock a little. He said he couldn't work the gearshift because he was shot in the hand. "Just use your other hand," I said, then turned the .50-cal back on the fighters in front of us. They were still sticking their guns just above the edge of the ground, doing a wild AK spray in our direction. My bullets couldn't penetrate the ground, so I was only suppressing them so they couldn't take accurate aim.

My buddy was still fumbling with the gears when I heard something off to our left. I looked over and saw another fighter was shooting at us. I swung the .50-cal around and sent one long burst in his direction. He fell over into the wadi. Either I hit him or he decided to run. I'm not sure which.

I could then hear other American trucks racing in our direction … but they were not there yet. Just then the driver threw the gear in reverse with his good hand and stomped on the gas pedal. We flew back over the berm and away from the enemy shooters while I suppressed them with the .50-cal, but then I was worried about backing into the helicopter that was still exploding. Rounds and missiles were popping off.

I shouted, "Stop! Don't get too close to that helicopter!" For a one-armed driver bleeding all over the place and in partial shock, my buddy did a great job. Even in that condition, he was still in the fight. Amazing.

I put the truck into park for him and got him out. I put pressure on the wound but it kept bleeding. So I then put a tourniquet on it with medium tension until the bleeding stopped, or at least slowed. Then I pulled out a pistol, gave it to him in his good hand, and told him that if anyone came over that berm to shoot them. I ran to get a medic. I grabbed my M4 and ran to the nearest truck that had just arrived and grabbed a medic that just happened to be there. He came with me, grabbed my buddy, and carried him back to his truck.

With my buddy being treated, I jumped into the driver's seat of my truck and drove back over the berm and in position to get back into the fight. I shut the truck down and climbed back up into the turret and started to suppress the enemy with the rest of my fellow Rangers who had by then arrived. Then I talked with the other guys about how to dislodge or kill those fighters down in the wadi and where their position was.

As I was firing, I saw something fly out from the gully, hit the ground and roll towards my truck, coming to a stop about ten meters away. "Oh, shit! Grenade!" I only could think of one thing to do: cover my nuts. I remember thinking, in a split second, that I didn't care if I got hit in the face, I just didn't want my balls to be blown off by a grenade. So I covered them up with my hands, closed my eyes, and looked down to cover my head with my helmet. Then the grenade exploded.

Nothing hit me. I was like, "Oh, whoa!" I looked around to see if there were any holes in me. None, except where a bullet had passed through my pants when we were being shot at earlier. For a moment I thought, "Wow, I got lucky," and then started shooting at the enemy again.

Finally one of our guys threw a grenade at them. That slowed down their rate of fire enough for two of our guys to run up there, pop into the wadi and shoot them both. That was the end of the firefight. After that, we set up another perimeter.

I later learned that the Apache was shooting its cannon and missiles at a Toyota truck with a mounted machine gun. It was sneaking up on us through all of those berms. The pilots had destroyed the truck and forced all of the enemy combatants to dismount and take positions behind the berm where I had found them.

Looking back, whether those pilots were dead or alive, I was at least going to get their bodies out of there for their families. As a Ranger, part of the creed that we live by is that we will never leave a fallen comrade behind. I would gladly do anything, and by any means, to get those pilots or my men out of there. As with any Ranger in the regiment, there is no hesitation to close on the enemy and destroy them in order to complete the mission, or help a soldier in need. This is why someone joins the regiment, and this is the essence of being in the regiment. I was just happy that I could be in the right place at the right time, and have the opportunity to help in any way that I could.

For saving the downed pilots, Olsen received the Silver Star, the third-highest military combat decoration for gallantry in action.

The day after Olsen's battle, CIA Director George Tenant appointed David Kay, an international weapons inspector, to head the Iraq Survey Group's efforts to locate evidence of Saddam's WMD program in Iraq.

Then, on July 22, coalition forces converged on a house in the northern Iraqi city of Mosul after a tipster reported it as the location of Uday and Qusay.

Our warfighters "swiftly moved into assault positions while infantry from the 101st Airborne Division set up a cordon around the villa to stop anyone from escaping," read one account of the raid. A bullhorn was used to tell the brothers to surrender. Their answer? A barrage of gunfire. At that point, "things just went ballistic," said one participant. "Those guys put up a massive fight." Using C-4 explosives, U.S. forces stormed through the iron front gate—the only viable entrance to the walled compound, participants said. From there, some began clearing the first floor, while others climbed back stairs and crossed the roof for other entry points."

After the attack, the lifeless bodies of Uday and Qusay were hauled away, ending at least one horrible chapter in Iraq's bloody history. Later that year their father would be pulled from a hole in the ground in a farming village near his hometown of Tikrit. After a lengthy trial, Saddam Hussein was hung three years later on December 30, 2006, in Baghdad. The hunt for HVTs would continue as former regime officials became part of the insurgency, and as the conflict bred new insurgent leaders who needed to be killed or captured.

**U.S. forces attack Uday and Qusay's hideout in Mosul,
July 2003 (*U.S. Army*)**

As for WMD, David Kay found no stockpiles and resigned from the Iraqi Survey Group in early 2004. He told Congress that he didn't think the stockpiles ever existed. However, during an appearance on Fox News Sunday in February 2004, Kay said, "Although I found no weapons, (Iraq) had tremendous capabilities in this area. A marketplace phenomena was about to occur ... sellers meeting buyers. And I think that would have been very dangerous if the war had not intervened."

Indeed. After more than a year of searches, investigations, and analysis, a final report covering the search for WMD was released by the Central Intelligence Agency in the fall of 2004. Among several key findings, the report stated that although Saddam didn't possess stockpiles of WMD at the beginning of the war, he certainly intended to possess them in the future.

"Saddam wanted to recreate Iraq's WMD capability ... but probably with a different mix of capabilities to that which previously existed," read the key findings in the CIA report. "Saddam aspired to develop a nuclear capability ... but he intended to focus on ballistic missiles and tactical chemical warfare capabilities." The report also stated that even though Saddam had no written plan or active WMD program, "his lieutenants understood WMD revival was his goal."

The ways and means to achieve that goal were kept close at hand, as well. Christopher Hitchens recounted one example in his memoir that he unearthed during a visit to Baghdad after the invasion: "Unnoticed by almost everybody, and unreported by most newspapers, Saddam Hussein's former chief physicist Dr. Mahdi Obeidi had waited until a few weeks after the fall of Baghdad to accost some American soldiers and invite them to excavate his back garden. There he showed them the components of a gas centrifuge—the crown jewels of uranium enrichment—along with a two-foot stack of blueprints. This burial had originally been ordered by Saddam's younger son Qusay, who had himself been in charge of the Ministry of Concealment, and had outlasted many visits by 'inspectors.'"

This wasn't an isolated case. More than ten years later, the New York Times published a story not only revealing that there were indeed thousands of WMDs found in Iraq—albeit in smaller numbers than the feared "stockpiles"—but also that some of our warfighters were exposed to chemical weapons. "From 2004 to 2011, American and American-trained Iraqi troops repeatedly encountered, and on at least six occasions were wounded by, chemical weapons remaining from years earlier in Saddam Hussein's rule," wrote C. J. Chivers in an October 2014 edition of the New York Times. "In all, American troops secretly reported finding roughly 5,000 chemical warheads, shells or aviation bombs."

Whatever conclusions one can draw from the WMD issue, one thing remains clear: the tyrant and his sons were dead, and their regime's goal—or fantasy, depending upon your estimate—of developing a WMD program was halted. The world was mercifully spared from a future where the Hussein family continued to possess unimaginable wealth and held aspirations for regional domination.

Nevertheless, while the initial invasion was successful and the regime was toppled, America's warfighters found themselves still fighting and taking casualties, especially in Baghdad and in much of the country's Sunni-dominated provinces. Some said the attacks were coming from the remnants of Iraq's army, while others, including many Australian officers who had experience fighting or studying that type of war, saw ominous warnings that an insurgency was sprouting from the ashes of Saddam's Iraq.

Meanwhile, in the supposedly friendly Shia south, a new threat to our warfighters was emerging in the ancient city of Najaf. The nation was spiraling towards the civil war the coalition was working to prevent, but simply couldn't stop.

CHAPTER FIVE

"IT WAS ALMOST LIKE FIGHTING ZOMBIES."

JUSTIN LeHEW
BRONZE STAR WITH VALOR, *NAJAF*

B*y early 2004, America's warfighters found themselves in a full-blown insurgency. This was no longer a question except in the minds of those unwilling to comprehend what was happening. To quell the problem, our generals had devised a strategy to train the Iraqi security forces, turn over responsibility to their commanders, and then decrease our troop presence, since our simply "being there" was thought to be the major reason for the fighting. That was only part of the reason, we'd later learn.*

Aside from general lawlessness and competition for power, there were growing political movements that were set against not only the coalition, but also the vision for a free Iraq that the coalition was trying to promote. These factions began to launch coordinated attacks against our warfighters, and each other. Roadside bombs and suicide bombers were their preferred method, and these tactics proved to be as costly for Iraqi civilians as they were disheartening for an American public who had thought, only months before, that we had won the war.

The attacks against political targets became bolder, as well. In mid-January 2004, a terrorist detonated a truck bomb at the gates of the "Green Zone," the area in Baghdad that housed the Coalition Provisional Authority and the various factions that comprised the Iraqi Governing Council (a collection of leaders who advised the coalition's political leadership). More than twenty were killed. Then, on February 1, more than fifty Iraqis were killed when two suicide bombers attacked the offices of the main Kurdish political parties in Erbil.

Reports of attacks against the emerging Iraqi security forces also became a daily occurrence. On February 10, at least fifty Iraqis were killed when a terrorist detonated a car bomb at a police station south of Baghdad, and the next day at least forty more were killed during a similar attack on an army recruitment center in Baghdad itself. Two days later, twenty-three Iraqis were killed during a firefight with insurgent forces who attacked a police station in Fallujah.

The indiscriminate attacks were vicious and often claimed the lives of many more innocent Iraqi civilians than Iraqi soldiers, policemen, or coalition forces. In one terrorist attack on a police station in the southern town of Basra in mid-April 2004, more than a dozen children were burned alive in their school bus that happened to be outside of a police station when the building was struck by a suicide bomber. Only one fifteen-year-old girl escaped. "I had just left the house," she told a reporter. "I opened the door and went out. I could see the bus. I found myself flying in the air and falling on the ground. I saw fire and smoke. It was a huge explosion. I couldn't get up again." The reporter described how the girl was shivering, shaking, and weeping, and that she said, "I can't believe all my friends have been killed. I'm the only one left."

These tactics horrified the American public, and our warfighters were seeing these atrocities occur daily all across Iraq, not only in the Sunni-dominated areas where former regime officials and al-Qaeda-linked militants were waging war, but also in the Shia areas. One of the principal Shia threats came from a little-known cleric from a prominent family: Muqtada al-Sadr.

"When U.S. forces rolled through Iraq in March and April 2003, most Shiites greeted them as liberators," read an article in the Fall 2004 edition of the Middle East Quarterly. "But, the demise of Saddam Hussein's regime in Iraq unleashed an array of forces that had been dormant or suppressed for more than three decades. From almost total political marginalization, the Iraqi Shiites found themselves at the center of political power. While some political parties ... participated in the U.S.-sponsored political process, it was

not long before U.S. forces became aware of a new force among the Shiites."

Coalition officials initially largely ignored al-Sadr, who was regarded as a thug by secular Shia leaders and as an uneducated and insignificant cleric by Shia imams. But the Shia youth found familiarity and purpose in his fiery sermons. "A large body of young, poor Shiites have found voice in Muqtada's violent populist movement," continued the article in the Middle East Quarterly. "In Muqtada they have found a leader who trumpets their rage."

While his reach extended throughout the Shia-dominated south and into Baghdad's sprawling "Sadr City" section, named for Muqtada's father, his base of operations remained in Najaf, about 120 miles south of the capital. It's a city of about half a million residents, mostly Shia, and it was founded not as a trading or agricultural community, as most cities are, but as a burial ground. Mohammed's son-in-law, Ali ibn Abi Talib, known as Imam Ali, was the fourth caliph and is the main historical reason for division between the Sunnis, the largest group within Islam, and those known as Shia, one of the smaller groups who believe Imam Ali was a true spiritual leader. Ali ruled from the city of Kufa, in modern-day Iraq, and as he lay dying from a poisoned sword attack, the imam ordered that his lifeless body be placed upon a camel's back that should be set free to wander in the desert. Wherever the camel first stopped, he told his followers that was where he should be buried. The beast walked for six miles before resting in the desert, and Imam Ali's followers immediately buried him.

From his desert tomb arose the holy city of Najaf and the sacred Imam Ali Mosque. Shiites believe that being buried near their beloved imam is spiritu-ally significant, so over the centuries millions have come to be buried in the cemetery adjacent to the mosque. The Wadi al-Salaam, or Valley of Peace, is a twelve-square-mile labyrinth of catacombs that makes up the largest cemetery in the world—and a uniquely morbid place to fight.

In his book, Battle for the City of the Dead, author Dick Camp de-scribes the cemetery well: "Scattered throughout the jumble of graves are vaults, each with two or three subfloors containing a line of crypts ... sometimes used as weapons-storage areas and hiding places ... Minarets one to two stories high tower over the cemetery, providing places of unhindered observation for snipers ... In places, the monuments are so closely placed that it is almost impossible to squeeze between them. Lieutenant Colonel Eugene N. Apicella described the terrain as 'a New Orleans cemetery on steroids.'"

The cemetery was a virtual hive of insurgents, protected from bombs and artillery by its haphazard construction and its religious and historic significance. If the coalition wanted to clear it out, it would have to be done by infantrymen moving not from house to house, but from crypt to crypt.

Within this environment, al-Sadr attempted to form a shadow government in opposition to the Coalition Provisional Authority, and especially the Iraqi Governing Council, whom he loathed as traitors. He also raised an army called the Jaysh al-Mahdi, or Mahdi Militia, named after the twelfth imam, whom many Muslims believe will return as Judgment Day nears. The militia's strength was hard to estimate, but it was thought to be anywhere between three thousand to ten thousand fighters. These black-clad insurgents were largely teenagers, unorganized and inexperienced, yet unified in their devotion to al-Sadr and their desire to kill coalition forces. Joost Hiltermann, director of a Jordan-based think tank that was studying the militias during that period, called the Mahdi Militia an "army of the dispossessed."

This "dispossessed" militia was first deployed in a major way in April of 2004 after the Coalition Provisional Authority, under the direction of Bremer, closed one of al-Sadr's newspapers in Baghdad. The action brought the fringe cleric to greater prominence. Many Shia were simply happy to see someone make a stand against the coalition, and the ranks of his militia swelled.

After several weeks of unproductive talks, the U.S. military attacked Muqtada's militiamen in Najaf, Karbala, and Kufa. Hundreds of the inexperienced fighters were killed. Even though they were our enemy, it was sad: teenagers, reportedly high on drugs, rushed into the open, spraying bullets at American forces who were forced to cut them down with precise and overwhelming fire. The militia eventually withdrew and al-Sadr accepted terms, publically telling his fighters to lay down their arms. But it was clear to some on the ground that he had privately told them to simply wait, and that another battle would eventually come.

Camp explained how al-Sadr never actually relaxed his grip on Najaf after the initial battle with coalition forces. "It was apparent to the observers that despite the cease-fire agreement, Najaf had not been turned over to the civilian authorities. Al-Sadr's Mahdi Militia still controlled the city by intimidating the population. His Sharia courts still existed, handing out sentences that ranged from mild reprimands to savage beatings and even executions."

Weeks after the ceasefire, on June 28, 2004, Bremer and the Coalition Provisional Authority left Iraq and handed control to a sovereign interim government headed by former exile Ayad Allawi. Coalition military forces remained and continued their mission of training Iraqi forces and providing security. While Iraq was reestablished as a sovereign nation and many hoped the violence would end, the war was still far from over. Baghdad, Fallujah, and Najaf were all slated to receive new contingents of forces. One of those units was Lieutenant Colonel John L. Mayer's Battalion Landing Team, 1st Battalion, 4th Marines (BLT 1/4). They had planned on being a reserve force in Baghdad, but as the U.S. Marine Corps History Division wrote, plans had changed. "Instead, it walked into the path of a raging storm, resulting in one of the most intense battles the Marine Corps had seen in Operation Iraqi Freedom. In the city of al-Najaf, the battalion engaged a fanatical enemy in a place where 'it rained shrapnel,' and machine gun, small arms, and rocket-propelled grenade fire reached intensities unknown to already battle-tested Marines."

One of those "battle-tested" Marines in BLT 1/4 was Justin LeHew. He was thirty-four years old now and his experience during the Battle of Nasiriyah—searching for the lost soldiers, being ambushed while crossing the bridge, and fighting in the city—not only earned him a growing reputation within the Corps, it taught him a great deal about war that he'd now pass along to younger Marines.

JUSTIN LEHEW: After the invasion of Iraq, I was back home for a few short months when I was promoted to first sergeant, which is the senior enlisted Marine in a company. The first sergeant primarily serves in an administrative and leadership role. He is responsible for the care and feeding of the Marines, for their families, and for making sure they're getting paid on time. Things like that, but he's also responsible for the morale, the discipline, and the efficiency of that unit and how it operates. He's the right-hand man of the company commander.

When I was promoted to first sergeant, I looked at the list of jobs that were there, and I knew what I wanted: to be on a Marine Expeditionary Unit and in the infantry. I saw that Charlie Company, First Battalion, Fourth Marines needed a first sergeant, so I packed my stuff, left the AmTrac world, and I went off to be an infantry first sergeant. I had 140 Marines under my charge, and I loved every minute of it.

A Marine Expeditionary Unit is about 2,500 Marines lashed up and put aboard naval ships with their own aircraft, artillery, and equipment. It's an entire air-ground task force. I was looking forward to it. You get to see different ports, and do different types of training. I was like, "Okay, this is going to be pretty neat."

When we got onboard the ship in May 2004, we got the word that we were actually going to go straight to Iraq. We only had a short time to train, so we started. It's amazing how the lessons learned from the first time in Iraq helped me prepare for the second time. I knew the heat we were going into. I knew the timeframe of the year when we were going in. It was going to be substantially hotter than when I was there the last time. I knew the area we were going into, in and around Najaf.

I remember reading about how Herb Brooks trained the 1980 U.S. Olympic hockey team, whose players were a lot smaller than the Russians, and certainly not as capable. Conditioning beat the Russians. Our team ended up outlasting them. When the Russians were too damn tired to defend their net, that's when the Americans turned on the gas. I remembered that lesson, and that's how we trained our Marines.

When we were onboard the ship riding over, we trained every day. Everybody wore their flak jackets and carried their weapons everywhere during our thirty-day transit. Wherever you went, even if you were in just physical training gear, you still had your boots on, no tennis shoes, and you still wore that flak jacket. For instance, one day the colonel said, "We're going on a ten-mile hike," and we did it on the ship, in full gear. Up and down the ladders. We were in full MOPP gear, too [*chemical weapons protective suits and gas masks—very hot*], so our lung capacity increased. That's how we trained. We conditioned those Marines to where there was nothing in Iraq that would deadline them outside of their own abilities. Their bodies wouldn't shut down.

That unit was as close to anything I've seen next to the "Band of Brothers" that we put into Bastogne in World War II. Every company that was in there supported each other. There was the rivalry you normally have, but everyone looked out for each other. The leadership looked out for the Marines. Nobody took from each other or hoarded for themselves. I had never felt like that before.

During our offload in Kuwait, we got the word that we were going to take over the battle space in Najaf. So we started studying parts of the city. To me, Najaf was just like Nasiriyah except Najaf was centered on a cultural heritage location, sort of like Mecca. Najaf is one of the holiest cities in the Shiite Muslim culture. It had a mosque with a big gold dome. Around that mosque was the cemetery, the largest cemetery in the entire world. So we offloaded our equipment and started the road march all the way up to Najaf. We linked up with a U.S. Army unit that had been in the area and did a left-seat-right-seat turnover, which is when a Soldier and a Marine do things together, so we could learn what they had been doing.

Was the weather any better in Najaf?

It was about 130–140 degrees. I don't even believe that myself when I say it, but we had Marines take photographs of the thermometers so people couldn't later say that we were embellishing. On average, I think the lowest it would go was about 120 degrees. Then, during the hottest portion of the day, around noon, it would climb back close to 140 degrees. Unimaginable. The heat complicated everything because we were so armored-up. We had twenty pounds of armor. Then we had our weapons, and as much water and ammunition that we could carry. But the conditioning we did was why I believe that we performed so well. We weren't shoving IVs in people's arms throughout the battle in Najaf like we had in Nasiriyah.

What did the Army tell you about Najaf?

They were pretty pissed because they hadn't gotten to mix it up that much with the enemy. Their hands were tied. I remember going into some of the early turnover meetings and seeing a map of Najaf with huge "exclusion zones" drawn in red. At that time I really didn't know what an "exclusion zone" was. One of them was the cemetery, and another was the town of Kufa, where Muqtada al-Sadr was from. He was the leader of the uprising in and around Najaf. I can remember Lieutenant Colonel John Mayer saying, "Well, hell. That's where the enemy's at." The Army knew that, too. I don't think those exclusions zones were drawn because the Army didn't want to go there. They were simply told by leadership, "You are not going into there." They knew, and they even told us, "This is where the enemy is."

But when the U.S. Army pulled out, the Marines were in full-scale combat inside of Najaf within weeks. After they left, we started actively patrolling in the city by putting U.S. troops in armored vehicles and on foot patrols to show our presence in the city. But we were still not allowed in those exclusion zones.

Then we started a secondary mission. We were tasked with training the 405th Iraqi National Guard Battalion that was supposed to be standing up to defend Najaf. So that became our principal mission in Charlie Company.

What did you think of the Iraqi recruits?

On any given day, half of those guys were fighting against us at night and then they came back and got training the next day. That was the weirdest feeling. Young Marines who were in charge of training those Iraqis, they were the ones that would tell you. They would come back to their gunnies or their first sergeant or their captain and say, "I've got a weird feeling about that guy." They came and got training, and the next day forty of them were missing. We just knew that we were training the enemy, and there was really nothing we could do about it. It was our mission.

Alpha Company had another training mission over in Diwaniyah, a few kilometers away, where they were training another Iraqi battalion. The remaining company, Bravo Company, was centered down in the city near a place called the Agricultural Center. They took over a huge hotel, probably nineteen stories tall, and from there the company controlled the center of the city.

Before the main battle started, we'd had a couple of firefights. Some of our patrols were shot at if they got too near Kufa. Some of our other patrols were getting into smaller firefights, too. Then it started to escalate.

On the morning of August 5th, we were at a compound in the city training the 405th. We heard that an Iraqi police station was being overrun. It was near a place called Revolutionary Circle, which was right across the street from that cemetery. Our Iraqi battalion was the unit that was supposed to handle the response. The Marine advisors and everybody else had to go with them, too.

That was the start of the fight in Najaf. When we were called in to reinforce that police station, that's when all the shit happened. What I mean by shit is that we had a "Black Hawk Down" scenario. An aircraft was shot down a few hundred meters off of Revolutionary Circle. When that aircraft was shot down, we had to secure the crash site. We were still in the firefight at the police station, too. More forces were getting sucked into the city, near the circle, and then all the sniper fire started coming out of the Wadi al-Salaam cemetery.

Sometime that afternoon Lieutenant Colonel Mayer got clearance to go into the cemetery. I can remember receiving the order at 1600 hours, near the end of the day. Our task was to push 1,500 meters into that cemetery, and it would be at night.

The cemetery has numerous entrances on its sides, but it also has a long road through its center. The reason that road is even there was, back in the day, Saddam Hussein was oppressing the people of that area so he decided, "I'm going to take a tank and bulldoze their graves to show them who's in charge." So he made this road through the cemetery. We called it the diagonal road because it ran diagonally through the cemetery. We knew that we had to hold that road, because it's the only way to get supplies in and out of the cemetery, and it was the only way to get casualties out, as well.

It's not like a cemetery in the United States. The Wadi al-Salaam is full of crypts and mausoleums. They bury a lot of them above ground in these little mud pits, and then the mud hardens up to brick-like statues. They are all interwoven, not like Arlington National Cemetery or anywhere else where there are distinct rows, or even a distinct elevation. In Najaf, it's random: large families, individual graves. Some have been there for hundreds of years, and some are freshly made. They were conducting around fifty burials a day inside that place, even throughout the fight.

Everything was underground. Each of those mausoleums— thousands of them—has subterranean levels. Not only were the fighters hiding inside them, they were stockpiling ammunitions and weapons, and U.S. military uniforms and Iraqi police uniforms. Some of the mausoleums would connect under the ground to other mausoleums. It was like we were standing atop a giant nest, or a huge ant farm. And the enemy knew where they were. They held all of the tactical positions, and they had command and control over all of them.

Soldiers search the mausoleums, tombs, and catacombs of the Najaf cemetery for weapons and enemy fighters, August 2004 (*U.S. Army*)

This was also the time when we were starting to learn about improvised explosive devices [*IEDs*]. They had wired different areas of that cemetery with IEDs, and also pre-plotted positions from the hotels surrounding the cemetery for snipers to have over-watch. They also had very good mortar men who really knew what they were doing.

So, late that afternoon, after the attack on the police station, we got some tanks and started pushing into the cemetery. We moved at a rapid rate of speed, and it gave me a sick feeling in my stomach because night was coming and we were moving quickly past areas that weren't cleared.

Then, all of the sudden, the Mahdi Militia started firing at us from the crypts, and they started shooting at us from behind, too. That's when I knew how bad of a situation that was. Once again, I don't know if it just follows me around, but it was like another case of the Battle of the Little Big Horn. We were completely surrounded. Except this time we didn't know where the enemy was. We couldn't see them. They would shoot at us, then jump into a mausoleum, and then come back out someplace else. They would hide behind the tombstones, too.

Then it started to get dark. We knew that we were going to lose track of where our men were, so we started getting people condensed closer together. I remember having a meeting with the company commander, Captain Matthew Morrissey, and the executive officer, 1st Lieutenant Larry Costa. We were lying on our stomachs, the fighting was going on all around us, and we were saying how we needed to keep that diagonal road really close. We couldn't push thousands of meters into the cemetery or we were going to lose our entire company. So we put up defenses near the diagonal road, and once we started doing that, the mortars started coming in. It was like they knew what they were doing. They started channeling us into certain areas to consolidate more Marines to hit with the mortars.

Then, around 2000 hours, it became pitch black except for the fire from the guns, and casualties started happening. We started hearing screams through the cemetery, screams for "corpsman!" Anytime that happened, I went to those spots. Then you would also start hearing screams, but you'd find out that a Marine wasn't shot; he twisted his ankle climbing across a mausoleum or he fell into a crypt. It was a hazardous place to move around in, and the darkness compounded everything.

From the official summary of action for the Bronze Star with Valor that LeHew would receive for the battle: "Upon making first contact with the enemy, Company C came under a hailstorm of RPG, machine gun, sniper, and mortar fire. Within minutes, thunderous explosions of 82mm mortars were landing within the company's lines wounding 3 Marines. First Sergeant LeHew left his position of cover and immediately went to the impact site to direct company corpsman to triage and assist the wounded. More mortar rounds impacted and he stood completely un-phased by the risks to his life. When complete, he moved approximately 500 meters to the company's combat trains staging point where on 3 separate occasions, he directed the fires of vehicle mounted weapons against enemy snipers and militia attempting to penetrate the rear lines of the company. For the next 6 hours, he repeatedly moved back and forth under the crack of sniper fire overhead and mortar explosions all around, to communicate the enemy situation and status of the Marines with the Company Commander. It must be made clear that even moving through the cemetery was extremely dangerous, filled as it was by thousands of crypts and tombs that offered concealment for enemy movement and under frequent accurate mortar fire. It is not an exaggeration to say that First Sergeant LeHew completely disregarded the risks, even as Marines were killed and wounded around him, and found his way to every hot-spot on the battlefield."

It was pitch black, so I had the Marines yelling to one another across the line to stay in contact. Every minute, I made sure that those Marines on the left and right talked to each other. Something like, "Hey, I'm up here, ten meters in front of you." "Okay, Jonathan, I got it. You're up there." We collectively accounted for everybody that way.

At one point during the night the company commander looked at me and said, "We need to figure out how bad this area is right now." So we came up with an idea. I said, "Hey, maybe we can get a crazy helicopter pilot to buzz the cemetery, at 200–400 feet, whatever they feel comfortable doing, because the Iraqis aren't going to let a target like that just go by. They're firing tracers, so if we could get them to shoot at the helicopter as it flies by, we'll figure out exactly where they're at by seeing their tracers fly into the air." So our forward air controller, Captain Randy Gibbons, a phenomenal Marine whose call sign was "Chimp," called in the air.

Chimp is a great human being, man. I've never seen anybody call-in air like that guy. If I had a son serving in the military, I would want him with those guys—the guys that I went into Najaf with, guys like that forward air controller. They were just that good. Chimp called in that gunship and when it buzzed that diagonal road it looked like the gunfight at the O.K. Corral. Tracers went at that helicopter from everywhere: behind us, in front of us, right next to us. One was twenty-five meters away from my position. Then the fighters all disappeared back down into those crypts. That's when I knew, once again, that we were completely surrounded.

We knew we needed to systematically deal with those crypts, so that's what we did. We started rolling fragmentation grenades inside the smaller ones, and Marines started to go down into the larger ones, just like those tunnel rats did in Vietnam.

I can't think of a more eerie place in the world to fight other than a cemetery. I simply cannot describe the eeriness of being there, under the cover of darkness and fighting in the largest cemetery in human existence. What made me sick to my stomach was the unbelievable reality that I was fighting in a paradox—I was fighting in a place that's dead. The entire place is dead. And, quite frankly, I could soon be one of the dead, as well.

On the first day of battle we had some superficial casualties. We had some people shot in the hand and some with fragmentation wounds. But they were back in the fight within a few days. We weren't losing very many.

The sun came up the next day, and it was hot as hell. Our Marines were running out of water because they were drinking it so damn fast. A major problem was that most resupply into the cemetery had to be carried by hand. We could only get vehicles on that one road, and from there all supplies had to be hand-carried into our positions within the cemetery. We would have to load Marines down with water bottles, putting them in their packs. Then they would hike for a mile into the cemetery, climbing over crypts, to get water to Marines. We tried to chill it, but of course by the time that you got there it'd be hot.

The tankers were overheating, too. What I mean by the tankers, I mean the actual men sitting inside the tanks were overheating. They had to come out, lean against the tracks, and we put in IVs of fluid to keep them in the fight because we couldn't lose those tanks out there. Once I saw a tank parked on the diagonal road, and about one hundred meters away from the tank was a young tank crewman with a M4 rifle. I ran across the street and asked him, "Hey, Marine, what are you doing?" He said, dazed, "I'm guarding the tank." I just found that to be the most comical thing: this eighteen-year-old crewman saying, "I'm guarding the tank." I started laughing. So then I said, "Hey, buddy, you need to come with me. One, you're out here by yourself. Two, let's go find out what's going on with the tank."

LeHew began bringing in Marines from his unit who were pulling guard duty at nearby forward operating bases. Elements from the other Marine companies joined the battle in the cemetery, along with several U.S. Army units. "It's gotta be big," LeHew said. The battled raged for several more days.

Numerous people got mortared and we had a couple killed. That's when we would hear the sickening screams of "corpsman!" You just knew that, at night, after the mortars came, you knew it wasn't going to be pleasant. The mortars would stop at night, and every morning religiously, at four in the morning, we would get woken up by the Mahdi Militia firing mortars. It was like clockwork every day. They stopped five times a day to pray. That was a very eerie feeling, too. You knew when it was going to come. Then the snipers showed up. They were mainly foreign fighters from places like Syria and Chechnya. We captured a few and learned that they were getting paid large amounts of money to come and kill Americans. They were very well trained.

Meanwhile, the average individual fighting for the Mahdi Militia was about as untrained as it got. For example, they would just hop on a bus in the city with their weapons, move through the streets to their battle positions, then they'd hop off with a mortar, fire two or three rounds—just harassing rounds without really aiming—and then they'd jump back on the bus and go somewhere else. There wasn't a cohesive unit, but we found out that they did have somewhat of a command and control infrastructure.

On August 6, a sniper hit a young Marine named Lance Corporal

Brooks right through the side of his armor, underneath his armpit. It was what we call a "sucking chest wound." I yelled for a corpsman, and then picked up Brooks. I'm only five-foot-eight-inches tall and weigh about 165 pounds, and I had Brooks in my arms, running out of the cemetery to the diagonal road.

I ran up to one of the Humvees. It had the windshields and the tires all shot out. I didn't know where the driver was, but I had our company clerk with me, a Marine by the name of Lance Corporal Edward Criss. Now, one of the things you need to know about your Marines is who they are when they're not wearing the uniform. I knew that Criss was a street racer in Los Angeles during his free time. He would always get in trouble with the cops. He was one of those "Fast and Furious" types. So I looked at Criss and said, "Get in and drive the Humvee!"

Criss whipped that Humvee around, with flat tires, faster than any car I've ever seen. We loaded Brooks inside and I lay across his body, plugging his chest wound. We were at least a mile down the far end of the diagonal road and we starting moving out. We were flying past crypts and mausoleums, and the Iraqis were firing at us the whole way. I knew that at the end of the road, just outside of the cemetery, there was a casualty collection point. But I also knew that we couldn't wait. Brooks needed to go straight back to our forward operating base. It was about fourteen kilometers from the middle of the city. So I told Criss, "Bypass the forward casualty collection point," because I knew that Brooks would have died if we stopped.

So on flat tires and a shot-out windshield, Criss tore out of the cemetery like a bat out of hell and drove through the city straight into the forward operating base. There was a helicopter triage team waiting and we offloaded Brooks and they took him away.

(LeHew later learned that Brooks survived. From LeHew's award: "Realizing the need to stabilize the Marine with no time to spare, his timely decision-making made the difference in saving this Marine's life.")

We were at the forward operating base, and I had always taught the Marines that if Humvees ever go to the rear, when they're emptied out, they don't ever come back to the front empty. So as I was dealing with Brooks, Criss ran over to the local mess hall and got the people to load up anything they could. Ammunition, as well. He used a bucket to wash out the blood and broken glass, and loaded the supplies. Within thirty minutes we turned around and drove right back into the center of that cemetery again, right back into the same firefight.

During the rest of the battle, whether we were firing or not, I would always do a headcount every thirty minutes. We would yell down from our positions and have the Marines respond back. I would checkmark that we got every one of them. But then, later that same day, August 6, we weren't getting a response back from Lance Corporal Larry Wells [*a twenty-two-year old Marine from Mount Harmon, Louisiana, a little north of New Orleans*].

During the previous two hours, a sniper who was located in a hotel about five hundred meters across from our battle position had been shooting at Marines who would expose themselves from behind the crypts. I can remember one of the sniper's targets, Lance Corporal Jason Stover. I was behind Stover by about twenty-five meters and I told him to stay behind a headstone. The headstone just barely covered Stover's body, and when that Marine would move, the sniper would take a piece of masonry right off the top of that headstone. I was trying to get Stover out of there, but this went on for a very long time. We would try to see where the shot came from, have our machine gunners light up that part of the hotel, and then get Stover to move a little before the next sniper round fired.

We went through that for about two hours. Then, 1300 rolled around, and that was about the time that the Mahdi Militia started to pray again. The clerics started howling their prayers from the minarets, and things got really quiet in the cemetery. I did my roll-call check, but we couldn't hear Larry Wells. A corpsman eventually found him about thirty feet from where he was supposed to be.

Larry had crawled behind a crypt during the firefight and took a sniper round right to his neck. By the time we found him, he was already long gone. I estimated that he was dead for about twenty minutes, judging by the blood around him and that it was already congealed. I knew, unfortunately, with that kind of a shot through his throat, Larry couldn't yell. He couldn't scream. He had probably just bled out instantly.

We had to get Larry out of there. We took the opportune time during the afternoon prayer to run some Marines—myself, Larry's platoon sergeant, Staff Sergeant Cisneros, and a couple other Marines—to retrieve his body. I knew, however, that moving Larry would probably be the most dangerous thing I had ever done. We knew there were still snipers in the hotels around the cemetery, and it would be a long, difficult process to get him out—carrying him over the mausoleums, across the uneven crypts, and exposed to the snipers during the entire time. So we put Larry on a stretcher and carried him out of the cemetery. Then, religiously, as soon as the clerics stopped the afternoon prayer, the cemetery erupted with gunfire.

(Larry's family learned of his death soon afterward. "I'm proud of my brother for everything he's done," his brother, Chad Wells, told WWL-TV of New Orleans.)

I still swear to this day that the sniper allowed us to remove Larry from the battlefield. I have no other explanation as to why we weren't all shot. He had been firing for two hours prior to that, moving from window to window. Whoever the sniper was, he knew his bullet killed that Marine. He knew what we were doing because he was watching us. But he didn't fire. We had Larry on a stretcher, with Marines walking at a good clip, up and down through the uneven parts of the cemetery, climbing on top of a crypt, laying the stretcher down, climbing off, and then handing the stretcher back down—all very slow, and all very open. There could have easily been five other dead Marines if that sniper had wanted to fire.

I don't know if the average person can relate to this. I've heard people say over and over again, "Those people are heathens. They have no values. They use bombs against kids." You know what? That is true, to an extent. The way we train our Marines, however, is that the person who is trying to kill you believes in whatever it is they believe in just as much as you believe in whatever it is that you believe in. Regardless if it's a uniformed fighter, or if it's an insurgent, it's man-to-man combat.

That's sort of how I think of that sniper. When that target's neutralized, whatever that individual did, it's, "I won. You lost." In the end, there is a mutual, hated respect between combatants.

So, we finally got Larry to a safe position and, I kid you not, as soon as we were behind a wall and put that Marine down, the prayer time ended and the sniper started shooting again, right at the headstones around us. That's why I know that he watched us the entire time. He knew that he had killed that Marine, and he knew that we were just trying to retrieve our dead. But once that action was complete, it was, "Now we're combatants again." I could probably spend my entire life trying to figure out why that happened. There's a higher power probably somewhere on both sides of the fence looking out for the beliefs of different individuals, I guess. At the time I didn't get wrapped up in why that happened. I was just thankful that it did.

Things went haywire inside the cemetery again. The snipers came back out, and fighting continued. We went through that every single day. We'd go in and out of the cemetery. We had numerous wounded; some people were shot multiple times. We also had several individuals who refused to leave the battlefield. Staff Sergeant Todd Boydston, the third platoon sergeant, is honestly one of the best infantry platoon sergeant in the U.S. Marine Corps. He took a round or some shrapnel through his thigh but refused to leave for a day and a half. The fragment was still in his leg. I had to order him, eventually, out of the battle, to go back and get that treated. But less than twenty-four hours later I saw him getting out of a vehicle that had returned from a supply run. Boydston was coming back to lead his Marines.

There were others like that, too. A Marine by the name of Corporal Martinez, who was a squad leader in Boydston's platoon, had fragmentation wounds in his back, but he wouldn't leave. There was also a tank platoon commander, Lieutenant Thomas. He opened the tank's hatch to look at something and took an AK-47 round in his arm. I saw him drop inside of the tank. I ran over and called inside, "Hey, you have to come out and get treated." He didn't want to go. We eventually got him out of the tank and he was medevac'd. Marines were getting hit left and right. They were all bandaging their wounds, and they were all continuing to stay in the fight. I never had one single U.S. Marine ask me to go to the rear. I never heard, "I need to go." They all had to be made to leave. They didn't want to leave their buddies.

This all went on for several more days until we got a little bit of a break. We got to refit, and started rotating platoons. Then on the 10th and the 11th we went after al-Sadr. We were going to do a hit on his residence and also a place called the Sadr School. When we tried to clear the school, we saw how the insurgents were hopped up on drugs. We started to find little packages of white pills all over the place. They'd pop these pills, get amped up on adrenaline or whatever, and then go fight the Marines. They would take numerous hits and just still keep going. It was almost like fighting zombies.

At the school, third platoon was making their way up a staircase. The insurgents at the top of the building started rolling Egyptian-made grenades down the staircase at them. A Marine took fragmentation wounds all to his front. We pulled up to a position near the school and I began to assess the situation. Lieutenant Seth Moulton *(now a congressman representing Massachusetts's 6th congressional district)* said that there were Marines in the building who couldn't get out because a sniper was covering the alleyway. He said that anytime they sent Marines down there, the sniper would shoot. It had held the Marines back, and they couldn't get their wounded out of the building.

I figured, at that point, probably the most expendable person in the company was the first sergeant. The other guys all had missions, but the first sergeant didn't. That's why I got to hop all over the battlefield. So I said, "Hey, I'm going to zigzag across this alleyway, and I'm going to try to get the sniper to shoot at me so you can see his position, and then take him out." In the alleyway, on one side we had the wounded Marine, and on the other we had a machine gun team. The idea was that I would run and try to get the sniper to shoot at me—hopefully miss—and the machine gunners would see where he was, and then kill him. That would also allow me to get into that building to find out what we were looking at.

I ran across the alleyway and thankfully the sniper missed me. Then our gunners pinned him down. I got into the building and saw the wounded Marine lying on the floor. A couple of other Marines were hit, too. I told the Marines that we'd throw a grenade, and when it went off, that was the cue for the machine gunners to pepper the far side of the building. Then we were going to hang on the far left side of the road and bring those casualties out. They said, "Roger that." I looked at the biggest Marine and told him to give his weapon to someone and carry his wounded brother on his back. So he did.

We threw the hand grenade into the courtyard. It went off, the machine guns started firing, and we hugged the wall on the left-hand side while the machine gunners were firing less than ten feet away from us. It worked. We ran the casualties out and eventually evacuated them out of there. About thirty minutes later we had an F-18 fly in and drop a five-hundred-pound bomb on the entire building. We were under fire restrictions during the campaign, which means that only a certain level of ammunition could be fired with certain authority inside an area. For a five-hundred-pound bomb to come down on a building, for instance, you had to get two-star-level approval from up near Baghdad. No kidding, it had already become that kind of war.

Justin LeHew leads a foot patrol across "sniper alley" in Najaf, August 2004 (*J. LeHew*)

After nearly three weeks of fighting, the decision was made to take down the Imam Ali Mosque that Sadr and all of the insurgents were hiding inside. They were actually firing weapons from inside the walls of that mosque, which is completely against the rules of war. So we were going to surround it and take them down. The plan was to link up with an Army unit and do a night insertion into the hotels that were on the outskirts of the cemetery.

At about 10:30 on the evening of August 25, we started rolling into the city. We came up on the far right-hand side of town. An Army Bradley sat on an embankment and started engaging targets, and then we started inserting Marine infantry all throughout the night, under fire. For the next three days the only type of sleep we got was maybe a ten-minute nod-off inside of a fighting hole or inside of a building, because from the 25th to the 28th, it was full scale Nasiriyah-type combat.

We inserted into a hotel on the outskirts of that mosque. When we went into the first level of the hotel, I instantly heard a call that we'd just lost a Marine, a lance corporal. This hotel once had an elevator shaft, but it didn't have an elevator anymore. At the bottom of the shaft was rebar sticking up. The Marines were operating inside the building with night vision devices but they had to get a feel for the darkness. The lance corporal stepped into what he thought was a room, but the room turned out not to have a floor. He was impaled on the rebar at the bottom of the elevator shaft.

When that happened, the insurgents who were in the top floors of the hotel started firing and rolling grenades down on the Marines. Then the insurgents who were in the basement started coming up, too. It was like a nest. So our second platoon started to go downstairs and one squad started chasing insurgents upstairs. I heard firing and screaming. It was close enough to where it was hand-to-hand combat in that building.

At the same time, another Marine squad positioned themselves to go downstairs. The lead person of that squad was a Marine by the name of Private First Class Ryan Cullenward. I was twenty feet behind him when all this was going down. As Ryan got to the top of that staircase—and he was using night vision goggles as well—and started going down, he ran head-first into an insurgent with an RPG who was running up the stairs. That guy was going to fire the RPG and wipe out the entire squad. Cullenward instantly slung his rifle behind his back and dove off that top step. He tackled the insurgent and wrestled with him down two flights of stairs. They wrestled for about thirty seconds in complete darkness in that basement.

A lot of Marine units these days don't train with bayonets. They leave them in the armories. They go and buy their own personal knives, for sure, but nobody really fights with a bayonet anymore. They don't hook them on the end of their weapons. But we had trained those Marines to use their bayonets. Ryan had the wherewithal to win that fight because we had rehearsed where everyone was going to have their gear, and he knew exactly where his bayonet was. Ryan grabbed his bayonet and stabbed that insurgent to death, and then he grabbed that RPG and brought it up out of that basement. We later used that RPG in the fight. Here's another thing: a reconnaissance platoon went down in that basement and found another insurgent hiding there. He hid in the corner while Ryan and the other insurgent fought ... coward.

We used that hotel as our operating base for the next three days. We did patrols in all the alleys, cleared houses, and did counter-sniper operations. There was a point when we were under so much sniper fire that Marines couldn't expose themselves outside of our building or they would get picked off. There was another unit next to us that I could hear over the radio, and a very tired gunnery sergeant called and said they were running low on ammo and had almost no food and water. You could hear it in his voice ... they were low on everything, and it would take a while for supplies to come from one of the forward operating bases. I walked over to my Marines and told them the situation, and told them to throw out one canteen and one magazine apiece. They did it without question. None of them were like, "What the hell is this for?" I told them their brothers in Alpha Company needed it to sustain their lives. We had Marines gather up whatever we could in blankets and, under sniper fire, we ran those supplies hundreds of yards over to Alpha Company's position.

Meanwhile, the insurgents who were killed during the past three days couldn't be removed from the battlefield. That got to be very unsanitary and a huge health concern to me. I went by a young machine gunner's position one day ... he'd taken his T-shirt, he'd ripped it up, soaked it in water and had wrapped it around his face. He was sitting in his gun position and had at least 100 flies on his face. They were from those bodies inside the building.

I couldn't simply carry those bodies out because the snipers would kill my boys, and we couldn't leave them there because it was a major health concern, so I requested twenty gallons of gasoline so we could burn the bodies. I called up my colonel, and he asked what the gasoline was for. When we said what we were going to use it for, he said, "Oh, for God's sake, don't burn them. Do something else. Find another way." We didn't get into an argument. He ran the whole battalion landing team, and his word was solid. But then again, he was also not at my fighting position. I had to do something.

The bodies had been decomposing for a couple of days, so we found some blankets to taco-wrap them in. No Marine wanted to do what we had to do. I got in the middle of it and tried to lead by example. I started picking up pieces of the dead and said, "This is what we have to do." A young Marine leaned over—he was about nineteen years old—and tried to pick up a body but the arm came off. He was just standing there, holding this arm, and the skin and everything else was falling off. That was one of those times I was like, "My God. You can never train for something like that." The Marine was in shock, and he started laughing. He wasn't laughing because it was funny, though. He was just in shock. I walked up to him and said, "Hey, here's what we're going to do. Put that arm in the blanket and roll it up." So that's what we did with the bodies.

Our plan was to then put all of the machine gunners on the roof to cover us from sniper fire while we ran the bodies across the street. We'd dump them in a house, get them out of the way, shut the door, and get back across the street. So we did that, exposing our Marines to sniper fire so we could properly deal with the bodies. I also told them to make sure their heads were facing in the direction of Mecca.

The Marines did it, and not one of them asked, "First Sergeant, why the hell would we do any of that?" We shot a direction with our compass, picked them up in the blankets, and rotated the bodies to face Mecca. Then we ran out, back across the street and into the hotel, all while under sniper fire.

Then the most disgusting thing happened. When the fighting died down late that night, the guys who were on watch heard wild dogs bust the doors down in those buildings and tear those bodies apart. When the sun came up the next day, I saw a dog casually walking down the street with a foot hanging out of its mouth ... right in front of my twenty-year-old Marines. How does someone get that vision out of his head? I still have it in my head, and I had a lot more maturity under my belt than most of those young Marines out there. They're going to have that image in their heads for the rest of their lives.

We were in full-scale combat on the 27th, and then somebody wakes up and says, "Hey, you can't shoot anybody today." A cease-fire had been brokered.

From the book Battle for the City of the Dead: "[Lieutenant Colonel John] Mayer was disappointed, to say the least. 'We had been slugging it out for three days ... and had lost Marines and soldiers. We felt the decisive blow would have gotten Sadr and all his henchmen.' Major [Glen] Butler was in agreement. 'Everyone I spoke with at the time was extremely frustrated and upset. We felt as though we'd delivered a steady pounding on Sadr's forces for three weeks and were about to deliver the final blow for a complete victory. But instead, Sadr and his thugs escaped.'"

Sadr didn't need to defeat our Marines to win the battle, at least in the eyes of his supporters. He only needed to survive.

We all came out from the hotel, and then we had to set up everything for the peace agreement. Civilians were walking through the city, and a staff sergeant saw a mass of people coming down the road toward us. There's CNN, Al Jazeera, and a bunch of other news networks, too. We put up a perimeter of Marines, and with an interpreter, learned that the individuals were there to retrieve the dead bodies. At first I'm like, "How would you know there are dead bodies here unless you were here too?"

They just had the look on them of pure hatred. "We're here to get our bodies," they said. So we took them to the house. When they went inside to where the bodies were, and saw that the dogs had eaten them, and what was left of them, they all went insane. They started screaming. They started trying to assault the Marines. They were throwing rocks. And then they started to tell the news agencies, "See what these barbarians did to our people!" We couldn't calm them down.

I got an interpreter and said, "You tell them this." So I told the story of what had happened, how the Marines risked their lives to remove the bodies of people who had only moments before tried to kill us, and they had the wherewithal to cover them up. Then I said, "Look at the direction their bodies are facing! Those Marines had the respect for your people to face their heads towards Mecca." When they heard that … well, all the yelling and screaming stopped. An ambulance came up and everybody calmly let them take the bodies.

There was one kid, maybe seventeen years old, who grabbed an interpreter and started pointing in a direction. The interpreter came over and said, "He would like to go over there. He says you're all missing a body." So we walked over and found a foreign fighter, lying behind some rocks with a rifle, less than one hundred meters away from that hotel. That seventeen-year-old kid was lying there, too, when the guy was killed. That's how he knew the body was there. But the look on that kid's face … it was sheer death. His eyes were sunk back in his head, just a thousand-yard stare. He looked like he wanted to kill everybody. He wasn't scared. So we helped load his friend in the ambulance, and they rolled out of there.

We didn't see much more after that.

The citation for the Bronze Star with Valor that LeHew received stated that his "combat leadership and judgment often meant the difference between life and death. On every occasion, under some of the most violent conditions imaginable, he performed his duties without fear. He is a warrior in every sense of the word, who has time and again demonstrated bravery, compassion, and esprit de corps beyond comparison."

Despite the terms of the cease-fire, Sadr and his militia remained armed and active in Najaf and other areas of Iraq. They continued to harass the coalition and attack Iraqi security forces for years to come. Camp ended his account of the battle, published in 2011, with a warning: "The political stalemate has allowed al-Sadr to emerge as a key power broker, and with the Iraqi government paralyzed, he may yet again rise to prominence. It is possible that sometime in the future, the United States may find itself across the bargaining table with Muqtada al-Sadr."

CHAPTER SIX

"IT'S NO FUN WHEN THE RABBIT HAS THE GUN."

AUBREY MCDADE
NAVY CROSS, *FALLUJAH*

I n late March of 2004, around the same time that al-Sadr began his first major insurrection after his newspaper was closed, four Americans who were driving through the predominantly Sunni city of Fallujah were ambushed and killed. Their bodies were mutilated, burned, and then hung from a bridge spanning the Euphrates River. While the victims are mainly remembered as being "contractors" working for the Blackwater security company, it's important to remember that they were all veterans of the United States Armed Forces. Scott Helvenston was a former U.S. Navy SEAL, Jerko Zovko served in the 82nd Airborne Division, Michael Teague earned a Bronze Star for his service in Afghanistan, and Retired Sergeant First Class Wesley Batalona was a U.S. Army Ranger.

The four experienced former warfighters were escorting a supply run that had taken a shortcut through Fallujah. They were ambushed by insurgents. Their two Mitsubishi sport utility vehicles offered no protection, and they had little time to return fire with whatever small arms they had. In his book, No True Glory, military author Bing West, who accompanied Marines into the eventual battle to retake the city, described the scene: "When an American with bullet wounds in his chest staggered out and fell to the ground, he was kicked, stomped, stabbed, and butchered. A boy ran up with a can of gasoline, doused the SUVs, and struck a match. The black smoke pointed like a finger up into the sky, attracting a swelling *crowd. Egged on by older men, boys dragged the smoldering corpses onto the pavement and beat the charred flesh with their flip-flops to show that Americans were scum under the soles of their shoes. A body was ripped apart, and a leg attached to a rope was tossed over a power line above the highway."*

American generals who were watching the incident unfold via over-head surveillance drones decided to wait until the crowd had exhausted itself before trying to enter Fallujah and retrieve the bodies. West continued: "The macabre carnival in Fallujah continued all day, crowds spurring on one another shouting, 'Viva mujahedeen! Long live the resistance!' Two of the charred corpses were dragged behind a car through the souk [a market]*, past rows of small shops and hundreds of cheering men, to the green trestle that the Americans called the Brooklyn Bridge. There the mob hung the bodies from an overhead girder, two black lumps dangling at the end of ropes."*

Their gruesome deaths and the manner in which their charred bodies were paraded before cameras evoked memories of the "Black Hawk Down" disaster in Somalia. Instead of calling for a withdrawal, however, Americans were now calling for revenge.

In early April, the deputy director for U.S. military operations, Briga-dier General Mark Kimmitt, said the response would be "overwhelming" and that coalition forces would "pacify" the city. Operation Vigilant Resolve, eventually known as the First Battle of Fallujah, was launched shortly afterward with the full weight of about two thousand Marines from the 1st Marine Expeditionary Force, supported by jet fighters and attack helicopters.

The Guardian newspaper described the battle as "the most ferocious urban street fighting in Iraq since the start of the war." Its reporters interviewed many residents who fled after the first days of the battle. "Four houses in my block were destroyed. The house behind mine was hit with two rockets," said one resident. "We didn't want to go out, but from my gate I could see fighters carrying their weapons in the street and fighting the Americans. I saw them being killed and I saw their bodies in the street." Another resident, a medical doctor, said, "I wanted to cry. It looked like a city of ghosts."

The battle was bloody, but the insurgents were besieged and on the run. Ambassador Bremer later recounted a video-teleconference at the time that ended with President Bush saying, "We need to be tougher than hell now. The American people want to know we're going after the bad guys. We need to get on the offensive and stay on the offensive."

Meanwhile, as the bodies piled up in Fallujah, the fragile coalition of Sunnis, Shia, and Kurds who comprised the Iraqi Governing Council were nearly in full revolt against the Coalition Provisional Authority. A respected senior member of the council said the attack was "collective punishment" of the civilians and threatened to resign. "You must call for an immediate cease-fire in Fallujah," Bremer was told by another council member.

"We were at the most critical crisis of the occupation," Bremer wrote in his book. "The stakes couldn't be higher. The Governing Council—which for better or worse had to help lead the country over the next crucial months—was on the verge of disintegrating due to Sunni resignation over Fallujah."

The coalition needed to find a way to accomplish the mission while keeping the Iraqi Governing Council together. After a few weeks of fighting, the battle came to a halt when negotiations allowed for the existence of a "Fallujah Brigade" that would, under the command of a former Sunni general, "police the city" and kill or capture insurgents or foreign fighters. The Marines withdrew, and the Fallujah Brigade entered the city. The short-term crisis over the city had ended but was eventually replaced by a longer-term problem. "The mafia has won and taken over there," one Shia member of the council warned Bremer. The member added that the brigade was "a move to Iraqi disunity and civil war."

In the weeks and months that followed, rather than fulfilling the terms of the cease-fire, Fallujah became a magnet for terrorists from across the Middle East. The de facto leader of the jihadist wing of the insurgency, a Jordanian named Abu Musab al-Zarqawi, had taken up residence in the city and was launching attacks daily. He proclaimed himself to be the "Emir of al-Qaeda's Operations in the Land of Mesopotamia." The city's leaders were stockpiling weapons and preparing for its second defense. Civilian aide workers were kidnapped from various locations in Iraq and taken to Fallujah, where many were tortured. Videos of Western civilians having their heads sawed off by insurgents became frequent, and any Iraqis found cooperating with coalition forces were treated far worse.

Something had to give. Sometime in September 2004, as American military deaths in Iraq crossed the thousand-mark, a senior U.S. official told ABC News that al-Zarqawi had amassed about five thousand foreign fighters in Fallujah and that killing him and his troops was now the "highest priority."

Once again, the Marines were called upon to lead another attack, this time code-named Operation Phantom Fury. The battle, which would eventually become known as the Second Battle of Fallujah, began in early November of 2004. Somewhere between ten thousand and fifteen thousand American warfighters, with their newly trained Iraqi counterparts, would attack the city.

In a New York Times article published about two weeks into the fight, war correspondent Dexter Filkins wrote that the battle was "the most sustained period of street-to-street fighting that Americans have encountered since the Vietnam War. The proximity gave the fighting a hellish intensity, with soldiers often close enough to look their enemies in the eyes." Filkins, who had covered the war since its beginning and had spent a considerable amount of time in war zones, said it was a "qualitatively different experience" than anything he had seen before. "From the first rockets vaulting out of the city as the Marines moved in, the noise and feel of the battle seemed altogether extraordinary; at other times, hardly real at all. The intimacy of combat, this plunge into urban warfare, was new to this generation of American soldiers, but it is a kind of fighting they will probably see again: a grinding struggle to root out guerrillas entrenched in a city, on streets marked in a language few American soldiers could comprehend."

Filkins was embedded with about 150 Marines from Bravo Company of the First Battalion, Eighth Marines (1/8 Marines). One of those men was a twenty-three-year-old sergeant from Houston, Texas, named Aubrey McDade. This was his second time in Iraq. "I went to Mosul, Iraq, in the previous year," McDade told me. "I saw some combat action there, but it wasn't nearly as kinetic as Fallujah."

AUBREY MCDADE: I wasn't supposed to go on the second deployment because I was injured with right ankle instability. I had to have ligament surgery to repair the damage. In fact, I was actually going to get out of the Marine Corps altogether.

Before the deployment, the rest of the company was at Fort A.P. Hill in Virginia doing pre-deployment training to get ready for Iraq. I was back at Camp Lejeune in North Carolina at the barracks just helping out, training new Marines who had just arrived from basic training. They were fresh out of the School of Infantry where you learn the basics of machine gun employment, but they hadn't had the opportunity to see how it really works in a field environment. Since we were all just sitting around the barracks, I decided to give them a little training. It was nothing serious, just little gun drills to get them familiar with how we did business in Bravo Company: respect for the gun, respect for the target. It was all machine-gunner stuff.

Then when the company came back from Fort A.P. Hill, we took over a basketball court for a few days and set up a scenario for training. We used the court as an operating base where we would house the Marines and then push patrols around the base to simulate street clearing. We didn't know what our mission was but we knew that we wouldn't operate as a conventional machine gun section. We trained as best as we could to be prepared for every scenario. We conducted mounted and dismounted patrols, set up defenses, vehicle checkpoints, stopped traffic, and searched vehicles.

Then I started thinking, "Damn, these new Marines are going to deploy and they haven't even been to the field yet, like the rest of the company." That just didn't sit right with me. I've always been my brother's keeper. I remember thinking, "I can't leave them hanging like this. So I'm going to go on the deployment."

I went to medical and I told them that I wanted to deploy. The doctor wasn't okay with signing off on me, though, because of my ankle injury. He said that he didn't want to send me to combat and have me become a liability versus an asset. I said, "Sir, this decision is one I need to make myself." I assured him that I wouldn't become a liability, and after being persistent about it, and maybe even a little bit aggressive, the doctor signed off on my paperwork. I was cleared for full duty. I went back to the basketball courts where my Marines were training and told them that I was going on the deployment. Man, they were ecstatic.

We arrived in Iraq on June 22, 2004, at al-Asad Airbase in the al-Anbar Province. We stayed there about a month getting acclimated, and then moved to the Haditha Dam, then to Ammo Supply Point Wolf, and then finally to Camp Fallujah. The camp was like a holding ground for everyone. It had Army, Navy, and Marines there. We stayed there for the duration of our time, from September, through the battle in November, and into January 2005.

Camp Fallujah was outside the city of Fallujah, give or take twenty minutes. Fallujah looked like a low-class California neighborhood. A bunch of small buildings, plaster everywhere, bars on the windows, big metal doors. Sometimes it was hot. Sometimes it was cold … like desert cold.

Fallujah was as much a trap as it was a city. Roughly measuring five-by-five kilometers, West wrote that the city "contained 39,000 buildings and almost 400,000 rooms, most offering solid protection against small-arms fire. The insurgents knew every alleyway and back door." He added that our military planners "believed the most deadly weapon would be the IED (improvised explosive device) in all its variations: buried under streets, rigged inside houses, taped to the sides of telephone poles, stuffed into manholes, hidden under loose lumber, or wired inside abandoned cars."

Before the battle, our days were routine; we'd get up, do a security patrol, and then come back. Then we would have alternate patrols that would go out every hour. During patrols we'd dismount from our vehicles to show our presence to the people, so they would know who we were and that we weren't there to destroy the population. We'd try to show them that we were there to help. We'd pass out chocolate bars and candy to the kids. We'd also gather intelligence about insurgents. But overall, patrol was a normal day-to-day thing. Other than that, we still trained hard so that everything else became second nature. While we were back at the camp, during whatever downtime we had, we'd try to have some type of normal socialization. We'd play cards and sports, like Wiffle ball or football. That was about it.

In the company, I was the machine gun section leader. We were the support element that provided heavy automatic weapons to the rifle platoons. When other Marines actually run into a structure to capture or kill whoever is inside, my team supports them with automatic machine gun fire from a distance. We cover their movements. As a machine gun section leader, I provided each platoon with two machine gun teams or one squad of machine guns.

The average Marine carries an M-16A4, which shoots a 5.56-millimeter bullet, but my machine guns shoot at a higher rate of fire and it carries a bigger round: 7.62 millimeter. The M-16 weighs about eight pounds, while the machine gun itself weighs 24.2 pounds, and the complete weapon system weighs upwards of sixty-five to seventy pounds. So, as section leader, I had three squad leaders, six team leaders, six gunners, and six ammo men—at least in a perfect world, but I didn't have that many Marines. So you do what you have to do. Sometimes I'd play section leader and sometimes I'd play squad leader.

When did you learn that your company would be attacking Fallujah?

There was a lot of word of mouth. Nothing solid, but we heard that we'd probably have to take the city, and how difficult it may be because the first time wasn't successful. I didn't want to go, though. Not because I was afraid or didn't think my Marines could handle it; I just didn't want anything to happen to my Marines. Then, one day at camp we heard what was going to happen, and then we started to prepare. We made sure that we held each other accountable and tried to think of every possible contingency. Then we just trained a lot: drill, drill, drill. All the time. Drill, drill, drill. That way, if a situation were to arise, our response would be second nature.

How did you feel about the coming battle?

At the end of the day, I think that we had to take Fallujah. We had to try to liberate that city, because, in my honest opinion, America doesn't support bullies, you know? Extremists were bullying people there. So we were called forward by the nation to go over there to right that wrong and put the bullies in their place. At the end of the day, I feel like we went over there for that reason … and for our brother to our left and the brother to our right.

We gave the city a warning. I cannot remember it precisely, but it was something like, "We're here to eliminate the insurgent presence. If you are not an insurgent or you do not want to be considered an insurgent, you have a certain amount of time to leave." I'm not trying to quote anybody, but for the most part that's what was said. We gave them about two weeks to leave the city. After we said that, it looked like a New York City traffic jam coming out of Fallujah. Everybody who stayed was determined to be an insurgent or a supporter of insurgents.

"The major ground assault began after dark on November 8 with the blinding flashes and monstrous claps of tank rounds, artillery shells, and mortar rounds, all bursting in blooms of lava red, while Basher (a U.S. Air Force C-130 gunship) hovered above the apex of the shells and pounded away" wrote West. "The insurgents responded with an unaimed barrage of rockets and mortars that traced arcs of red sparks across the night sky. It was raining intermittently, with a cold wind gusting from the east. Throughout the city, the few residents remaining huddled behind thick cement walls."

As McDade and his Marines entered the city, they heard a Psychological Operations unit (soldiers who use information-based products to influence the enemy) using their large truck-mounted loudspeakers to blast a song by the heavy metal band Drowning Pool. "Let the bodies hit the floor! Let the bodies hit the floor!" screamed over the loudspeakers, announcing the coming fury.

A few days later, on November 11th—Veterans Day—McDade was deep in the city, and he was about to do something that earned him praise from the highest ranks within his beloved Corps.

We were attached to first platoon and I was acting machine gun squad leader. I had two machine gun teams, and we were clearing the city. They were using tunnels to escape. They would barricade entrances and exits in the city to channel us into one-way-in, one-way-out traffic. It was challenging because when going house-to-house we tried to maneuver as well as the enemy, but they knew every little hiding spot because they were in their own house.

Earlier that day, Major General Richard Natonski, who commanded the 1st Marine Division, told reporters that the assault was "ahead of schedule" and that our warfighters were clearing the city house by house, and building by building. Along with finding a suspected "slaughterhouse" where coalition forces believed hostages like Nick Berg were tortured and decapitated, they kept finding weapons and fighters hidden in nearly every mosque and school they inspected. "This is the enemy that we fight," Natonski said. "He does not respect the religious mosques or the children's schools."

As the Marines pushed through the city, they found shocking evidence of atrocities committed by al-Zarqawi's followers. Houses that seemed outfitted for torture, with human-sized cages in rooms covered in blood. Nearly naked bodies of Iraqi men, presumably those who wouldn't fight the coalition forces, were found bound and shot dead in ditches. As the Marines were walking down one street, they came upon one particularly horrible sight. "The body of a blonde-haired woman with her legs and arms cut off and throat slit was found Sunday lying on a street in Fallujah," read an article from Agence France-Presse. One of the agency's photographers travelling with the Marines noted that "the woman was wearing a blue dress and her face was completely disfigured." Our warfighters saw atrocities like that on a routine basis, yet they had to press onward, and fight.

One of the challenges our warfighters faced was that their enemy could seemingly take a near-inhuman amount of gunshots before falling. This was because, as with al-Sadr's teenaged Shia fighters in Najaf, al-Zarqawi's Sunni fighters in Fallujah were dosing themselves with high levels of drugs. "The guys we're shooting are so high, that they don't even realize that they've been shot," wrote an Army soldier named Matthew Epps in his autobiography about his role in the battle. "To solve the problem, our dismounts start aiming low; shooting the enemy in their legs, the idea is that no matter how high they are if we shoot their legs out from under them, then they will go down where we can finish them off at our convenience. It works."

It was close to midnight that day, maybe ten or eleven o'clock. It was dark, but the moonlight wasn't that bad. Some places in the city still had power, so there were even a few streetlights. We didn't need our night vision goggles at the moment, but we still had them. It was fairly quiet, too. I heard gunshots in the distance, but I didn't know where it was coming from. We were moving to our next checkpoint, which was a pretty tall and solid building. We were going to take it, establish a foothold until daylight, and then wait until we got our next objective. On the way to the checkpoint we saw what looked like Iraqi Army soldiers.

Filkins, who was traveling with McDade's company at the time, described what happened next in his New York Times article: "On one particularly grim night, a group of Marines from Bravo Company's First Platoon turned a corner in the darkness and headed up an alley. As they did so, they came across men dressed in uniforms worn by the Iraqi National Guard. The uniforms were so perfect that they even carried pieces of red tape and white, the signal agreed upon to assure American soldiers that any Iraqis dressed that way would be friendly; the others could be killed. The Marines, spotting the red and white tape, waved, and the men in Iraqi uniforms opened fire. One American, Corporal Nathan Anderson, died instantly. One of the wounded men, Pfc. Andrew Russell, lay in the road, screaming from a nearly severed leg."

I was near the rear of the platoon and set up my machine gun teams so they could provide supporting fire and over-watch to the platoon, and to cover all avenues of approach to their rear. I heard what was happening, the shooting and the screaming, so I turned my machine gun teams over to Corporal Justin Rose, a gun team leader. I rushed forward and found my platoon sergeant, Staff Sergeant Eric Brown. He had already established a casualty collection point in the courtyard of a house.

I asked him, "What was going on, staff sergeant?" He explained how they had been ambushed, that he thought Corporal Anderson was dead, and that two Marines were injured and stuck in the alley. It was too dangerous to get them because there was very little room to maneuver, few places to hide and protect yourself, and the enemy had full aim down the alleyway. The enemy could see, and then shoot, anyone who went to help. You'd have maybe five or ten seconds before you were shot, too. I thought for a second, and then I told my platoon sergeant, "I can go get them."

What made you say that?

I was a set of free hands. I was just a squad leader. I had a lot of confidence in my machine gun team and I knew they could hold down the avenues of approach and provide security without me. I knew that I had an ability to help out. I also was one of the Marines that had been in Bravo Company 1/8 the longest. I got to the unit on April 22, 2000, and this battle was in November of 2004. Usually, Marines rotate units every three years, so every Marine in Bravo Company I knew very well. I had a hand in helping to train and mold them into the Marines they were. I had a lot of personal time and interest invested in every Marine in that platoon. We were really tight. They had respect for me, and I had respect for them. The Marines who were injured in the alley were my Marines! So I wasn't thinking about what could possibly happen to me. I was only thinking about what could I do to potentially help them.

The platoon sergeant said, "If something happens to you, I won't be able to get to you either, because our tank support is about ten minutes out."

I told him, "That's fine. Just don't let me die."

He said he'd do the best he could, and to be careful.

At the time, I weighed about 190 pounds, but with all my gear I probably weighed in excess of 250 pounds. Now, I'm not the fastest man in the world but I'm pretty quick. I thought that it would be harder to shoot a fast-moving target than a slow-moving target, but that gear would slow me down. Besides, the vests can stop 7.62 caliber bullets, but I wasn't entirely sure about everything they had to fire down the alley. Maybe they had something heavier. So, in the end, I felt that moving faster would be better than maybe being able to stop a bullet. I left my helmet on, but took off my vest, all of my rounds and gear, and left my weapon inside the casualty collection point.

I remember standing on the edge of the courtyard talking to Staff Sergeant Brown and waiting for a lull in the gunfire. The enemy had that alley lit up and there were very few pauses or breaks in the fire. Lance Corporal Douglas Kulbis had come up to my position asking how he could help. I told him just to stay low and move to a position that would allow him to engage the enemy.

So once I heard the gunfire slow down a just little bit, I dashed across the street and Kulbis went about his way. I was able to dive behind a porch where I saw Lance Corporal Robert Kelly. He was pinned down by the fire, too, and I asked him, "Hey, where the Marines at?" *(Kelly, who told McDade where to find the wounded Marines, would later become a second lieutenant. He was killed while leading his platoon on a foot patrol in Afghanistan in 2010, and is buried in Arlington National Cemetery.)* As I was talking to him, bullets were hitting the front side of the porch. I remember thinking, "How am I going to be able to get out and get over to those Marines?"

Around that time, one of my machine gun teams started firing at the building with the insurgents. I heard his rounds firing, and then I heard the insurgent rounds stop hitting the porch. He was providing good covering fire, and gained the enemy's attention. So, I jumped up and ran down the alley. I can't remember how far, but it wasn't that long of a distance.

The first Marine I found was Private First Class Russell. He was screaming. He told me he had been hit. I looked down and saw that his leg was barely hanging on. It was pretty bad. So I took all his gear off, picked him up in a fireman's carry, and started running. As I was running bullets were impacting all around my feet and I heard Russell say, "Sergeant, please don't let me die."

This all happened so fast, but I could tell the insurgent's shots were getting closer. He was bracketing his shots in front of me, all around me, and soon he'd zero-in on me directly. I kept running and then, all of a sudden, I stepped in a pothole and fell. Russell and I hit the ground just as the insurgent fired a burst of rounds that struck the ground at precisely the location we'd have been running if I hadn't stepped in that pothole.

That hole saved our lives.

I picked Russell up again, this time in a sort of baby hold. I finally got him across the street and into the casualty collection point. Our corpsmen, Doc Naderman and Doc Davis, were pretty young and shocked but they still did a remarkable job at patching him up and stopping the bleeding.

I went back into the alley, this time low-crawling. I don't know if the enemy had night vision capability or not, but I think they could see me. Rounds were impacting on the ground all around me. I made my way to some bushes for cover, but the tops of the bushes caught fire from all of the ammunition being shot my way.

I found the next Marine, Lance Corporal Carlos Domenech, near a porch. He had been shot between the neck and shoulder. It went clean through. He was a little Marine so he wasn't too hard to carry, but I still took off all of his gear to make it lighter. He was adamant about keeping his weapon, though. So I picked him up in the fireman's carry and started running down the alley. As I was running, Domenech was shooting. I'm not sure if he even knew what he was shooting at, but he was.

I ran maybe twenty or thirty meters with Domenech and got him to the casualty collection point just as the tank arrived. A tank has a phone on the back with about a fifty-foot cord so you can pull it out, get away from the tank, and then talk to the Marines inside. Obviously, my adrenaline was pumping and I wasn't paying attention because the tank's exhaust burned my face and hands. Anyway, I knew the approximate direction of the enemy—they were inside a building facing the alleyway—and so I communicated their location to the tankers.

They fired two rounds in the direction of the building. It was probably the loudest thing I have ever heard. Again, I didn't take the cord out far enough so I was right behind the tank when it fired. It felt like I had a concussion. I was dizzy; everything went dark. When I gathered myself together and was able to orient myself again, I went into the alley with Staff Sergeant Kenauth, a combat cameraman, and recovered Corporal Anderson. Whatever round hit Anderson, they said he was killed on impact. He was already dead.

Anderson was buried a few days later in North Bend Cemetery in Danville, Ohio. In his obituary published in the Akron Beacon Journal, his family wrote that he had wanted to be a Marine since he was ten years old. "His family recalls the pride that Nathan displayed as a result of serving his country and his drive to be the best Marine and son he could be," read the obituary. "Nathan once told his family that a 'thank you' coming from another human being that did not even share the same language, made it all worthwhile." Nate was well liked by all those in his company. McDade and his fellow Marines took his death hard.

You know that old saying, "It's no fun when the rabbit has the gun"? That's how I felt. We were whipping ass and taking names, but then we lost some of our own. It really set in with the Marines. The most difficult part of combat, for me, was being the guy who didn't and couldn't show weakness. I had to figure out a way to motivate my Marines to stay in the fight. I remember saying, "Hey, let's get this shit done and get payback!"

So I starting running around and yelling at Marines. "Pick your heads up! The mission's not over with! We have a job to do! Let's go get those bastards!"

Once our wounded were medevac'd out of there, we picked up, moved to our checkpoint, and continued the mission. [*Pauses for a moment.*] Wow, that was crazy.

Aubrey McDade (second from left) prepares to patrol through Fallujah,
November 2004 (*U.S. Marine Corps*)

*Nearly two weeks into the fight, on November 19, Lieutenant General John
Sattler told Pentagon reporters that the battle had been won. "We feel right
now that we have ... broken the back of the insurgency, and we have taken
away this safe haven," Sattler said. He added that the battle smashed al-
Zarqawi's network and "scattered" its remnants. "I personally believe, across
the country, this is going to make it very hard for them to operate. And I'm
hoping that we'll continue to breathe down their neck." Not all agreed with
that assessment, however, with some Pentagon officials saying privately that
the insurgency had "shown itself to be an adaptable band of dedicated killers"
that will remain a problem for years to come.*

*A few days later, the New York Times carried an article detailing how
al-Zarqawi recorded a message to his followers after the battle in which he
blamed lack of religious support for the defeat. "You have let us down in
the darkest circumstances and handed us over to the enemy," al-Zarqawi
said. "You have stopped supporting the mujahedeen. Hundreds of thousands of the nation's sons are being slaughtered at the hands of the infidels
because of your silence."*

*In his book about the effort to re-take Fallujah, West criticized the
initial withdrawal in the spring of 2004 that eventually necessitated a
second battle. "The singular lesson from Fallujah is clear: when you send*

our soldiers into battle, let them finish the fight. Ordering the Marines to attack, then calling them off, then dithering, then sending them back in constituted a flawed set of strategic decisions. American soldiers are not political bargaining chips."

Meanwhile, as al-Zarqawi tried to reconstitute his network of fighters elsewhere, McDade and his fellow Marines from Bravo Company 1/8 returned home in early 2005 ... and mourned their dead.

Filkins later travelled to North Carolina to attend a memorial service for those who didn't return. "I'd been expecting a ceremony and a parade and a band, with a lot of American flags and a cheering crowd," Filkins wrote in his book about the war. "As it happened the ceremony was held inside a gymnasium on the base ... about a third full, mostly with members of the battalion and their girlfriends and wives. There wasn't anybody there from the town, as far as I could tell, no reporters from the local paper and no band. About half of the gym's bleachers remained tucked in place against the walls."

The scene observed by Filkins was a stark reminder that we had sent our military to war, not our nation. Our country continued to live like nothing had really changed in our lives, while our military fought, bled, and died in foreign deserts.

Filkins continued, describing how he sat in the bleachers scanning the rows for familiar faces of the men whom he'd walked the streets of Fallujah with the previous November. "I spotted a guy seated in the front row. They'd put him on the floor so he wouldn't have to climb the bleachers. A metal brace encased his leg. The brace was so large and it stuck out so far that it resembled the scaffolding on a building. Or a birdcage. It belonged to Andrew Russell, whose screams I'd heard in the alley that night. He was moving very slowly, but the leg was his."

McDade was at the memorial service as well, but his mind kept drifting back to that battle.

When we came back to the United States, all I wanted to do was go back to Iraq. I had a personal vendetta. As a professional, I probably shouldn't have been that way. But I was a young Marine and I was thirsty for revenge. I really was. I kept asking to go back.

One day my first sergeant took me aside and said, "I've been there. You're pissed off, but it's not a good move for you to go back over there right now. You've done your job in Iraq; let someone else do their part. You're married. You need to do something for your family." So he sent me to be a drill instructor in Parris Island, South Carolina. Looking back, that probably wasn't the best idea, either. I was fresh off the battlefield and wanted to train recruits to be combat-effective instead of basically trained Marines. I was a hard knock, and wanted to push them to my standard, not necessarily the Marine Corps standard.

One day, I was incentive training the recruits, what we call "I.T.'ing," and the depot commander, Brigadier General Lefebvre, walked in my squad bay.

Was it unusual for the depot commander to come to your squad bay?

Oh, hell yeah. If the depot commander comes on your deck, it's either real good or real bad. He was a brigadier general and I was a sergeant. He outranked me by light-years. I called everyone to attention, saluted, and reported the deck. I was wondering if I was doing anything wrong, and then Brigadier General Lefebvre said, "Carry on."

I thought, "Well, I'm not doing anything wrong, so I'm going to keep on doing what I'm doing." Then the general said, "Drill instructor, do you really think you should be doing that while I'm on deck?" I thought it was a trick question. I've always thought that if you're doing the right thing, keep on doing it. I was within the parameters of the SOP [*Standard Operating Procedures*] so I said, "Yes, sir," then started I.T.-ing the recruits again.

Brigadier General Lefebvre said, "Drill instructor, get in the house."

The "house" is where the drill instructor stays in a squad bay. I said, "Yes, sir," then sent the recruits away. I walked into the house with the general. He said, "Sergeant McDade, have a seat. Do you know why I'm here?" I said, "No, sir." Then he said, "I came to see where you want me to present you your Navy Cross."

It stunned me. I said, "Excuse me, sir?" The general said, "Yeah, where do you want me to present you your Navy Cross?"

"Sir, I don't have a Navy Cross," I said. "Yes you do," he said. "Are you Aubrey Lean McDade, Junior?" Then he read my Social Security Number and showed me the citation.

"Yes, sir. That's me," I said. "But I don't want it." Brigadier General Lefebvre said, "Well, son, I'm sorry, but you've got it. It's right here in my hand." I said, "Sir, you can take it back to wherever you got it from. I don't want it."

Why didn't you want to accept it?

People who know me would never know that I had these experiences because I don't talk about them. It's nothing that I feel people need to know. And I also felt that an award is something to celebrate. How could I possibly celebrate us having to go through that shit and my Marines dying? You can't celebrate that. I knew that people wouldn't understand. They'd be congratulating me. I didn't want that. I would probably punch them in the face. That's how I was at the time. I was young, twenty-four years old, give or take, and I just wanted to bury it. I didn't want to let that be a part of me.

Then Brigadier General Lefebvre said, "Son, I know you're humble, but think of this as not an award for just you. It has your name on it, but you just have the honor of wearing it for the Marines you served with." Once he broke it down like that, it put me at peace. Now I'm more mature. I'm a man of strong convictions. I now understand that having this award gives me an opportunity and a voice to let people know what an amazing set of men I was able to serve with and lead into combat, because a lot of people don't know their stories. Now I have the opportunity to tell them.

So I told Brigadier General Lefebvre that I wanted to be presented with the Navy Cross in front of my old company, Bravo Company 1/8, and in front of my recruits who I was training. So he sent buses to Camp Lejeune, North Carolina, and picked up the guys from my old company and brought them to the graduation ceremony for my recruits at Parris Island, South Carolina. General Lefebvre pinned the Navy Cross on me during the ceremony. It was very nice.

How was it to be with the members of your old platoon?

It was bittersweet. Everyone talks about the deployment itself, but regardless of being shot at, regardless of how many people's lives I've taken, or how many of my Marines were killed, the hardest part about that whole deployment was coming home. I hadn't ever had a chance to release. So after the ceremony we went downtown and just ran amok. I don't drink, but I did that night. We had a good time and I was able to release. I was able to cry. I was able to laugh, and then cry some more. It was great. It was really great. We were a tight-knit family. We had to endure adversity, and that made a strong bond between us. I haven't seen them in years, and I really miss them.

How was life at home?

I'm not going to lie; I had a pretty rough time after coming back from Iraq. I was angry all of the time and was easily irritated. In that kind of personal situation some people binge eat. Some people turn to drugs. Some turn to alcohol. Some lift weights. I sought refuge in my Marines. I dove into the work and had a pretty rough time in my marriage. I don't know if I had a death wish or whatever, but I was really reckless. I'm just glad that my woman was strong enough to stay with me through it all. I was a terrible husband! Now, everything is good. I'm still married to the same woman.

Tell me about meeting the president.

Yeah, that was pretty cool. After I received the Navy Cross and my recruits graduated, President Bush and Mrs. Laura Bush invited my wife and I to the State of the Union Address. We got a tour of the White House, got a White House coin and some cuff links, and stayed in a top-notch hotel.

We watched the speech and as we were leaving the White House afterwards, President Bush saw me and said, "Marine Corps! My favorite branch." He shook my hand and asked, "Where are you from?" I said, "I'm from Texas, sir." He said, "Texas! My favorite state. You a Cowboys fan?" I said, "No, sir."

He said, "Well, two out of three, you can't beat that." He shook my hand and then pulled me aside. "So, Marine," he said. "What are they saying on the streets?"

I said, "Sir, they're talking about how you're over there for personal reasons. You're over there to benefit the United States. You're over there trying to get their oil, or that you're trying to do something that your father should have finished during Desert Storm."

He said, "Well, Marine, you're a straight shooter, huh?" I said, "Yes, sir. I don't hold any punches. You ask me and I'll tell you." He said, "I can respect that." Then the president said, "I want you to deliver a message for me. Can you do it?" I said, "Yes sir."

Then President Bush said, "I want you to tell our people that I understand their reservations, but you also tell them this: If they're willing, I'm willing. I'll pull all of our troops out of Iraq if they're ready to get their rifles and their guns and fight this battle on American soil. Because this fight has to be fought. It might be fought here or it will be fought there, but it's going to be fought. No one will ever attack our nation, create the tragedy and chaos like they did, and it go unanswered. That will never happen again. Can you pass it?"

I said, "Yes, sir. I will." He shook my hand again, gave me the presidential coin, and walked me to the doors that everyone else had left from.

Years later, during President Bush's farewell address to the nation from the East Room of the White House on January 15, 2009, he mentioned the fellow Texan and "straight shooter" he met in the hallway years before. "I have confidence in the promise of America because I know the character of its people," Bush said, recalling that he saw our nation's character "in Staff Sergeant Aubrey McDade, who charged into an ambush in Iraq" and rescued his fellow Marines.

CHAPTER SEVEN

"THEY PICKED THE WRONG DUDES TO MESS WITH."

JARION HALBISEN-GIBBS

ARMY COMMENDATION MEDAL WITH VALOR, *BAGHDAD*

O*ne of the long-term aims of the White House's strategy was to establish a true democracy in Iraq so that its sectarian squabbles could be settled by popular consent through a constitutional process, while providing an example to others throughout the troubled Middle East.*

"The rise of a free and self-governing Iraq will deny terrorists a base of operation, discredit their narrow ideology, and give momentum to reformers across the region," President Bush said during a May 2004 speech at the U.S. Army War College. "This will be a decisive blow to terrorism at the heart of its power, and a victory for the security of America and the civilized world."

The president went on to describe his administration's five steps to achieve that freedom: hand Iraqis back their government, establish security, rebuild the nation's infrastructure, bring other nations into the effort, and, finally, have a national election. It would be no easy task, as our military and diplomatic corps knew all too well.

Bush added, "Helping construct a stable democracy after decades of dictatorship is a massive undertaking. Yet we have a great advantage. Whenever people are given a choice in the matter, they prefer lives of freedom to lives of fear. Our enemies in Iraq are good at filling hospitals, but they do not build any. They can incite men to murder and suicide, but they cannot inspire men to live, and hope, and add to the progress of their country. The terrorists' only influence is violence, and their only agenda is death. Our agenda, in contrast, is freedom and independence, security and prosperity for the Iraqi people. And by removing a source of terrorist violence and instability in the Middle East, we also make our own country more secure."

The goal of a democratic Iraq has remained a highly criticized objective, mainly because it would have to introduce, or force, a foreign concept of government onto an ancient tribal culture. Many people said that Americans couldn't hand out democracy like baskets of fruit and expect it to work well. Iraqis would reject a secular democracy, some said, and without a realistic alternative in place, chaos would reign.

Some experts believed that only a benevolent strongman could successfully rule a place as tumultuous as Iraq. Perhaps, but for a strongman to have real authority and the means to keep it, he has to seize power, not be granted his office by a foreign military. Nobody had done that. Meanwhile, the ability of each faction to grow in strength during the occupation assured that no single individual could assume total control. They had to work together, and only a democracy provided the process through which that stood a chance of happening.

Other experts believed that Iraq should have been partitioned. That is, divided into three nations comprised of its largest parts: Sunni, Shia, and Kurd. But this was just as problematic because, despite ethnicities being clear majorities in some areas, Iraq's provinces aren't neatly and completely comprised of only one group or another. There are sizable minorities in each province, and the cities are also diverse, especially Baghdad. Oil reserves and religious shrines important to each group are scattered across the nation, some deeply within another's proposed partition.

Those who supported a partition never fully explained what they thought would happen to the Sunnis who lived in the Shia-dominated south, or the millions of Shia who lived in majority Sunni Baghdad, or the Sunnis who lived in the Kurdish-dominated northern provinces. They also never fully explained what would have happened to the Christians and the other minorities, or who would get the oil revenue, how pilgrims would visit sacred sites, and how families would be compensated for lands *and businesses lost during the partition. Most experts on the region didn't believe we could expect people who had lived in a town for centuries to simply load up their cars and relocate to parts of the country they may have never even visited. A river of refugees would have flowed across whatever lines were drawn. Those reluctant to leave would have been harassed or killed, and many believed civil war would have ensued, followed by decades of war between the sides that lost lands rich in oil or those of deep religious and tribal significance.*

Some contended civil war would come no matter what, which shows there wasn't a "right" way out of the dilemma. With a host of bad and worse options, many thought that only a democracy could give the Iraqis a chance to keep their nation from devolving into a land of ceaselessly warring provinces. For better or worse, democracy was Iraq's best chance at internal peace. So in early 2005, millions of Iraqis went to the polls.

"Defying death threats, mortars and suicide bombers, Iraqis turned out in great numbers on Sunday to vote in this country's first free elections in 50 years, offering a powerful, if uneven, endorsement of democratic rule," read a January 30, 2005, article in the New York Times. "In some polling centers, the mood turned joyous, with Iraqis celebrating their newfound democratic freedoms in street parties that, until the elections, were virtually unknown in this war-ravaged land."

Although most Sunnis boycotted the election, its relative success was a blow to the insurgency, but not enough to dishearten its fighters or slow the growth of its distinct factions. The insurgency was, after all, more of a group of rivals loosely united against a common enemy. As noted by Lionel Beehner in a May 2005 paper for the Council on Foreign Relations, "the insurgency comprises several groups of fighters … Baathists, foreign jihadis, and Iraqi nationalists … united by their desire to disrupt the political process and drive U.S. forces out of Iraq. But each element of the insurgency is also driven by its own unique motivations." Beehner added that those motivations included a complete Baathist or a complete Islamic government, basic nationalism, tribal squab-bles, and the lesser but strong elements of crime and revenge. Then there was the growing involvement of outside forces, notably Iran.

While the insurgency was comprised of several groups, one of the most dangerous remained al-Zarqawi's collection of Islamic extremists, many of whom came from outside of Iraq. As the State Department's report on terrorism put it, Iraq had become "a melting pot for jihadists from around the world, a training ground and an indoctrination center." Contrary to popular belief, these fighters weren't largely disillusioned and unemployed youth who could be easily persuaded to leave the battlefield by promises of social acceptance or a steady job. One study by the Global Research in International Affairs Center found that the incoming jihadists "came from wealthy or upper middle class families. Some were students who left their studies in order to join the battle in Iraq." During the six months prior to the release of the study, the center also noted that 61 percent of all foreign fighters killed in Iraq were from oil-rich Saudi Arabia, and of these, "some were highly educated, and the list contained several professionals, including two businessmen."

Tributes to the fighters were regularly published on websites attributed to al-Zarqawi's group. One such entry was made in honor of a Saudi named Abu Anas Thuhami, who killed himself in a suicide attack during January's elections: "He chose to be with the virgins of paradise ... He used to talk frequently about the virgins of paradise and their beauty, and he wished to drink a sip from the sustenance of paradise while a virgin beauty wiped his mouth."

There's simply no outreach initiative or employment program that could deter someone like Thuhami and others like him. They came to Iraq to fight and die, because a martyr's death was the reward. The only way to stop them was to kill them first. Two years into our involvement in Iraq, al-Zarqawi's death cult had become our primary enemy. Debates about the necessity of the 2003 invasion were now largely academic and could only hope to inform future decisions. Iraq had become the central battlefield in our war against al-Qaeda and other Islamic extremist groups, and it was a fight we couldn't walk away from—and one we could not afford to lose.

"We are taking the fight to the enemy abroad so we do not have to face them here at home," President Bush said during the U.S. Naval Academy's commencement ceremony on May 27, 2005.

A few hours before the president spoke those words, one American warfighter showed exactly what "taking the fight to the enemy" looked like when his convoy was ambushed in Baghdad. He was a newly minted

Green Beret named Sergeant Jarion Halbisen-Gibbs from Ann Arbor, Michigan. As the new guy on the team, his Special Forces teammates teased him for being a "sandal fighter" because of his martial arts background and slapped him with the unfortunate nickname of "Jar Jar" soon after his arrival. "Yeah, like Jar Jar Binks, that dorky character from Star Wars," he told me. "Oh, how I hated that name." But another interesting thing about this warrior was his profession before becoming a Green Beret.

JARION HALBISEN-GIBBS: Believe it or not, I was a cook. I worked in a few different restaurants, but mostly at Zingerman's Delicatessen in Ann Arbor, Michigan. It's still around and they get damn good reviews. I didn't have a college degree and was just working to pay the bills and make enough money to pay my dojo fees. I was, and still am, an avid martial artist, and I earned my black belt right before I joined the military.

Why martial arts?

I don't know why people are drawn to certain things. From a very young age, I'd be swinging sticks at trees and fighting with little homemade staffs. Thankfully, my mom noticed my interest and put me in a karate class when I was only five years old. I took to it like a duck to water.

Looking back, it's now clear that I needed a positive outlet for aggression. I liked learning how to fight, which is a great skill to have. You get athletic ability from the effort and you also feel like you've done something significant or accomplished something unique. Now, learning to fight is truly a lifelong pursuit because nobody ever becomes the perfect warrior. You don't get your black belt and say, "Okay, well, now I am a badass; I don't have to train anymore." What your black belt actually means is that you know how to throw a punch and you know how to throw a kick. You can always become better than you are now, both physically and mentally. I guess that's how I view the martial arts; it allows you to train your body and your mind. I've practiced many styles of martial arts over the years, and probably the best way of describing them all is that they develop the mind-body concept of a warrior.

Where were you on September 11th?

I was twenty years old and living in Ann Arbor. I had worked a shift at the restaurant the night before and my mom called and woke me up that morning. She said, "Honey, you need to put on the news, now." I turned on the television and couldn't believe what was happening. My first thought was, "Holy shit, I guess we're going to war."

I went over to my buddy Nick's house and I sat there all day watching the news with him, his brothers, and their dad, Randall. It was heartbreaking. I remember sitting there and hearing Randall tell us all, "Boys, you're about to realize how much you don't know about the world."

After that day I was sure that war was coming, but like so many others I was expecting it to be something brief like Desert Storm. America would go in with the military forces it already had, kick a bunch of ass, and then pull out. I didn't want to join the military, go through the months of training, and then be late to the party, getting stuck on some base doing nothing but training for the rest of my enlistment. I think many other young men felt the same, which is why you probably didn't see long lines at the recruiter's office that next morning. Boy, were we wrong.

So I waited and I watched, and after several months it eventually became clear to me that America was going to be at war for the foreseeable future. So, I enlisted in the summer of 2002.

Had you given it much thought?

Joining the military had always been in the back of my mind. I dreamed about being in the military when I was a little kid, watching movies and television shows about war. I was always very interested in those. There were veterans around me when I was a kid, too, and anytime I had a chance to sneak up and listen to Vietnam vets share war stories—the kind kids weren't supposed to hear—I was absolutely enthralled.

A buddy of mine's father was a Vietnam vet. I still remember that he had a scar from when a Viet Cong soldier stabbed him in the hand. I would ask him questions about the war and about military life in general. He was a wise man and one day, after seeing that I was developing a sincere interest in the military, he looked at me in a very solemn way, and I felt he was about to impart something serious. He said, "Son, I wouldn't change my time in the Army if I could, but that experience is also something you cannot take back. Please remember that."

After several years in the Army, and several combat deployments under my belt, those words make a great deal of sense to me now.

Why did you join the Special Forces?

I'd never given much thought to the Special Forces, to be honest. I was only thinking about becoming an Army Ranger. I had heard stories about what the Rangers did in the Vietnam War, and how they were always out in the jungle fighting. So from an early age I associated joining the Army with joining the Rangers. I knew those guys were always damn busy, and that's still true to this day—the actual Ranger battalions, not just the guys who have been through Ranger School, but the "Batt' Boys," as we call them. They are good dudes and they do great work.

So that day in the summer of 2002, when I sat down in front of the recruiter's desk, I had a singular focus. Looking back, I was also probably a bit too eager. "I want a Ranger contract and want to go right now," I told him. "I have my stuff in the car, too. Let's go."

The recruiter laughed. "Slow down, turbo," he said. "It doesn't work quite like that." So we spent a little while talking about my background, my education, and my work history. Then, after hearing about my experience working at the deli and some other restaurants, the recruiter tried to talk me into being a cook.

"Hell no," I said. "There's no way I'm joining the army to be a cook."

I enjoyed working at Zingerman's Deli, and I know many dedicated soldiers who are 92 Golfs, the designation for food service specialists—army cooks. They have to do everything regular soldiers do, except at the beginning, middle, and end of the day, and sometimes even during the middle of the night, the army cooks have to feed the rest of us hungry, grumpy bastards. And if there's one thing soldiers have been bitching about since the beginning of time, it's the chow. So no, I didn't want to be an army cook, period. If I was going to join in the military, I was going to be a full-time warrior. I was going to fight. So I quickly convinced the recruiter that he was wasting his time trying to steer me into anything but combat arms.

"Rangers," I said again. "I'm already a cook, and don't want to be one in the army. I want to be a Ranger."

The recruiter got the message, but after checking the openings he said, "Sorry, I don't have a slot open for the Rangers right now." To say I was disappointed would be an understatement, but then he said something that changed my life forever. "But what I do have is an 18 X-Ray contract. Have you ever thought about Special Forces?"

I was surprised. "Oh, yeah," I said. "Of course, but those guys are ninjas. Don't you have to be in the Army for a few years before you even get a shot at trying out for Special Forces?"

"That's how it used to be," he said. "Now they need more guys."

I learned that the 18 X-Rays were kids that would join Special Forces directly. They would take a lot of tests, and if they scored high enough, they got the chance to try out for Special Forces training right off the bat. That was a huge leg up over everybody in the regular Army because, before September 11th, soldiers who were already in the military had to pay their dues in an infantry or an airborne unit before earning the right to try out for Special Forces. But since the attacks, the Army needed more operational guys, so the leadership decided that if a guy off the streets scores high enough on the entrance tests, and he makes it through the training, the Special Forces Qualifying Course, or Q-Course, just like everyone else, then he's earned the privilege of being part of the Special Forces.

I then asked, "What happens if I don't pass the Q-Course?"

He said, "Then you go into the regular Army. By the time you get there, you would have already gone through basic training and jump school, so you'll probably go into one of the airborne divisions."

I said, "Okay, so, if I pass, I get to join the Special Forces. If I don't pass, I get to join an airborne infantry unit? That sounds like a good deal. I'll give it a shot."

Worst-case scenario, whatever happened during the training, I was still going to go and fight for my country. That's what I wanted to do more than anything else. So I joined, and I was the fifth 18 X-Ray to come from Michigan.

Special Forces instruction is legendarily intense. In some years nearly 80 percent of all 18 X-Ray candidates fail to complete the training.

But before entering the Q-Course, candidates must first pass an assessment phase that's designed to weed out all those who aren't physically and mentally fit enough. In his book, The Guerilla Factory, former Special Forces colonel and course instructor Tony Schwalm described many of the challenges awaiting candidates. In what he remembered as "the most significant psychological challenge" of his own assessment, Schwalm wrote about the day he and other candidates were taken to an ice-covered lake about two hundred yards wide. The thin ice was broken and the soldiers were given a daunting task: construct a raft from rucksacks and ponchos and float an injured man across the water while keeping him dry. They made their raft, then stripped-down to avoid getting their own clothing wet, and swam the distance in about twenty minutes. "As our bodies became submerged in the water, my stomach muscles went into spasm and began contracting uncontrollably," he wrote. "I clenched my jaw and tried to stop shaking."

Even those who are able to pass such grueling tests often wash out in the Q-Course as the program becomes increasingly difficult. Soldiers spend many months, often longer than two calendar years, completing several rigorous courses covering military strategy and tactics, advanced combat skills, unconventional warfare, survival, advanced foreign language training, and many other skills. What emerges is a special kind of super soldier, with the knowledge to handle just about any challenge and the iron will to never stop. "Sometimes we succeed. Sometimes we fail," Schwalm wrote. "But we will die trying."

My training was just over two years. I know that sounds like a long time, but I was very lucky because I didn't have a lot of breaks between the courses and I didn't get recycled back to another class during any portion of the training. It was one class right after another. I entered basic training on October 23, 2002, and signed into the 10th Special Forces Group on November 16, 2004.

Tell me about the Green Berets.

The reason we're called Green Berets is simply because President John F. Kennedy authorized Special Forces guys to wear the green beret to distinguish them from the rest of the army. The term stuck after that, and we still wear the green berets in garrisons to this day. It's a pretty distinguishing feature, and the guys definitely wear it with pride.

The term "Special Forces," which is synonymous with the U.S. Army's Green Berets, is often confused with the term "Special Operations," which is the giant umbrella for all of the special operators in the entire military—the Green Berets, the Rangers, the Navy SEALs, the Combat Controllers from the air force, and some of the other unspoken-of entities. So when a military person says "Special Forces," or "SF" for short, they usually specifically mean the Green Berets. When someone says "Special Ops," they're talking about all of the special operators out there, including the Green Berets.

There are a handful of Special Forces Groups, which are the units we belong to, and each group has a specific area in the world that they are entrusted with training for and operating in. Those areas change depending on the conflicts, but at the time, 10th Special Forces Group, which is the group that I was attached to, was helping out 5th Special Forces Group with Iraq.

Individually, I joined the team as an 18 Bravo, which is the designator for a Special Forces Weapons Sergeant. An 18 Bravo is the team's subject-matter expert on everything from pistols to sniper rifles to crew-served machine guns, and also mortar systems and rocket systems. If you can shoot it, an 18 Bravo knows how.

On the team, a major portion of our capability is the overall tactics we use to accomplish the mission. Tactics-wise, it's not just a twelve-man Operational Detachment Alpha Team that goes out and kicks in doors, finds, and then finishes bad guys. Green Berets do a lot more than just that, much more than simply direct action missions. We mostly do what's known as unconventional warfare. In the broad spectrum of conflict, unconventional warfare takes a multitude of different forms. The most highly publicized form is direct action—kicking-in-the-doors missions—because that's visually badass; guys riding around on helicopters, kicking doors down, getting in gunfights. But there's a lot more to Special Forces than just that. There's a very large human piece to it, which is working with partner nation forces, also known as foreign internal defense. That's working with the country's indigenous military and/or police forces to empower them to be able to conduct their own operations and promote stability within their own country. We train, and then advise, foreign forces to fight their own battles.

What's an Operational Detachment Alpha?

An Operational Detachment Alpha, or ODA for short, is also known as an A-Team. It's a twelve-man team, which in the grand scheme of things doesn't sound like a large force. But when an A-Team is well trained and well equipped, they can become a very powerful combat multiplier.

For instance, when an ODA teams up with the right people, organizes them, and then teaches them to fight, that single twelve-man team can field a battalion-size force, which could be about a thousand fighters.

That's a good option for our commanders, who often find themselves unable to deploy a large number of American troops into an area for whatever reason. So instead of having to send a platoon, company, or battalion of American soldiers into an area—with all of the support required and consequences such a deployment can bring—they can send twelve Green Berets instead. And if things go right, they end up with the same amount of combat power, or maybe even more.

It's been said that a Green Beret is like a jack-of-all-trades but a master of none because it only takes twelve of us to train up such a large contingent of fighters across multiple fighting skills. But I like to think of us as more like Play-Doh; we can fit nicely into a round hole or square hole because we're so adaptable and our skills are applicable to many different types of missions.

On an ODA, you've got a captain who is the team's leader, an 18 Alpha. He will have a warrant officer, a 180 Alpha, who helps him in his leadership role. Then you have the team sergeant, or "Team Daddy" as he's known. He's a master sergeant and an 18 Zulu. He's the senior non-commissioned officer and his role is similar to that of a platoon sergeant or a company first sergeant. The Team Daddy takes care of the boys, focuses on their training, and makes sure they're competent in their fields.

Then you have all of the worker bees. Those guys have the good jobs, in my opinion. You have two 18 Bravos, a junior and senior. Those are the weapons guys, which is what I was. You have two 18 Charlies, also a junior and senior, which are the demolition and construction guys. It's funny, they build stuff and then they blow it up. Then you'll have two communication guys, 18 Echos, and then two medics, 18 Deltas. Those guys spend almost eighteen months learning how to be Special Forces medics, learning everything from trauma to sports injuries to treating different illnesses that we encounter in some of the third-world countries we deploy to. So those medics are a very, very important part of the team.

Why did you want to become a weapons sergeant?

I actually wanted to be an 18 Charlie, one of the construction and demolition guys. I'm not sure how or why the leadership makes the decisions they do, but after they told me that I was going to be an 18 Bravo, a weapons guy, instead, I was more than satisfied with that assignment because I grew up around guns.

I was raised on a farm for a long period of time. I'd shot guns and had always liked them, but once it became my profession it also became one of my passions. I'm a bona fide gun nut, and am also a Level 1 sniper. I'm really into long guns and everything sniper-related. But during training it was more of a hurdle. To get through that portion of the course, you have to learn everything you can to become a competent weapon's sergeant because your teammates are going to rely on you for that expertise. You have to know the job inside and out, forwards and backwards.

Jarion Halbisen-Gibbs in the gun turret (*J. Halbisen-Gibbs*)

During the time I was going through the Q-Course, the wars in Afghanistan and Iraq were raging, so there was definitely a sense of pride in where we were, and what we were doing, but there was also a sense of urgency in learning. We wanted to learn everything we could and bring that to a team, and then into the war. We were taught that you were either going to be an asset or a liability to your team, and none of us ever wanted to be a liability.

So, after all of that training, when I got to my Special Forces Group I was still thought of as a rookie. Everyone who showed up was thought of the same way. Even if you had regular Army experience before that, the teams still think of you as a rookie for at least the first year. Honestly, that's a pretty fun year because you get to know the guys on the team. You start establishing a reputation, and your reputation is one of the most important things you have as a Green Beret. You want to be known as a good dude. Some of the qualities that epitomize a Green Beret are being determined, competent, generous, and respected, and you can't just start that overnight. You have to earn it every day you're with the team, right from the beginning.

I was very fortunate to have been assigned to my team. I was in Bravo Company, 3rd Battalion, 10th Special Forces Group, Operational Detachment Alpha 083. It was a combat dive team, which is pretty much the same thing that SEALs do [*laughs*]: underwater infiltration. So, that skill doesn't really have much to do with fighting in the deserts of Iraq or the mountains of Afghanistan. The only diving you'll be doing there is diving out of the way of a bullet. But if you've ever heard of the CDQC [*combat diver qualification course*], or seen the special on the Discovery Channel, it's hands down one of the toughest schools in the entire military. It takes people out of their comfort zone in a way that no other course can do. For instance, you can take somebody who's a stud on the land. Maybe he can run a ten-minute two-mile, do a million push-ups, and be an absolute specimen of strength and endurance. But when you put him in the water, if he's not comfortable in that environment and if he has even the slightest challenge making decisions under pressure and keeping calm in stressful situations, then he's not going to pass. Being on a Special Forces dive team helped me because it breeds a form of mental toughness that cannot be achieved in other schools, and helps you grow your ability to make calm decisions under intense pressure. This pays dividends down the road, especially when you end up in combat.

So tell me about your first combat deployment.

I was twenty-three years old and serving as the junior weapons sergeant, an E-5, on Operational Detachment Alpha 083. We had been in Iraq for about six months and I had gotten into quite a few firefights, but they weren't heavy or significant, in my opinion. I know it may sound contradictory, but they were "safe" firefights. Either we were returning fire from a fortified position or we knew the threat we were going after. There hadn't been too many surprises or concerns, at least from my perspective.

Those "safe" firefights remained the extent of my war experience until late May of 2005, and then my experience changed for good. It was around 1 p.m. on May 29, and I was in a small convoy of two armored Humvees making our way down a busy boulevard in Iraq known as Route Force. That was a large, four-lane thoroughfare divided by a median, with one-lane access roads on each side that were lined with buildings and shops of all sizes and sorts. To give you an idea, think of a busy road in a major city back home, the road with all of the shops, grocery stores, and restaurants, and then squeeze everything together in a massive clutter, string power lines and telephone cables in every direction, paint everything gray, and then throw twenty-five years of war, crime, corruption, and poverty on top of it all, and that's Route Force.

I was the gunner in the first vehicle, sitting in the turret and scanning for anything that needed to be shot, basically. But instead of having the standard machine gun that's mounted on most vehicles, I had a Dillon Aero minigun, which most people would say resembled a compact Gatling gun. Miniguns are mostly mounted on the sides of helicopters, but we retrofitted one for our Humvee. It wasn't normal, but then again, we're not usually in normal situations either.

The minigun is an awesome weapon: a six-barreled, electronically-fed machine gun that shoots 7.62 bullets at a rate of fifty-two rounds per second, or two thousand to three thousand rounds per minute, depending upon which firing button was pushed. We had thousands of rounds stored in the back of the truck, too, with a belt feeding them straight into the gun. It was awesome to shoot a minigun in training, but for a gun enthusiast like me, shooting the minigun in combat was like a musician playing Jimi Hendrix's personal Stratocaster. It's phenomenal.

Now, one might think that such a powerful six-barreled gun would have an equally powerful recoil kick; it didn't. You could shoot the thing one-handed. Others might also think its rate of fire meant it was inaccurate and that the bullets would fly everywhere; it wasn't. Because of the lack of a kick combined with the rate of fire, you could see where your rounds were impacting during real-time, and then just walk the rounds onto the target. It was almost like having a laser sight made out of bullets. So you could be quite accurate while throwing thousands of rounds downrange, which makes the minigun a perfect weapon for defensive fire ... and for chasing bad guys down alleyways, as I'll explain in a moment.

Behind the minigun (*J. Halbisen-Gibbs*)

There was only one other person in my Humvee, and that was the driver, another Green Beret who I cannot name due to security reasons. He was a very experienced, combat-focused mentor, and I was lucky to have him there. He's a great dude, and an awesome driver. Now, being a turret gunner is fun, but most of the time gunners aren't shooting anything. They're mostly just hanging on for dear life while whatever vehicle they're in barrels down a crowded road or around some sharp curve. It's a rough ride. The driver, on the other hand, is the guy getting all of the action. Whether you were in a military vehicle, or in one of the civilian cars or trucks we impounded and used for missions, driving around that city was absolutely crazy. We called it the Baghdad 500. Hell, I'd pay to go back and be able to drive around like that one more time.

In the convoy's second and last vehicle, which was about 150 meters behind us, were the team's captain, a Navy SEAL, and another Special Forces teammate named Mike Lindsay, who happened to be with me during another major firefight in 2007. We were all headed to one of Saddam's royal palaces where we were meeting a few pilots to plan and coordinate the following day's mission. We weren't expecting anything to happen during the drive, but later we learned that our convoy had inadvertently driven right into a major insurgent operation.

Nearby, an insurgent group was about to attack one of our major detention facilities in an attempt to free some of their worst members. It was a classic jailbreak. Their plan was simple: the primary attacking force would assemble near the detention facility and wait while a secondary force created a diversion by setting up an ambush along Route Force. Their idea was to draw attention—and forces—to the main road, and then the primary force would attack the detention facility and free the bad guys.

So we're rolling down Route Force, oblivious to the coming jailbreak, of course. From my position in the turret I could see that we were coming up on a rather large-sized, concrete pedestrian footbridge that spanned from one side of the highway to another. Two Humvees were parked under the bridge for some reason, maybe to check passing vehicles for contraband or something. I'm not sure. A couple of Strykers, which are a cross between an armored personnel carrier and a small tank, were parked on the other side of the road.

We didn't know what they were doing there and saw no reason to stop. There were Humvees and Strykers operating throughout the city, so the sight of them along the highway wasn't unusual. The driver didn't let up on the gas and we kept moving down the road, soon to pass under the footbridge and continue onward to the palace to coordinate our next mission.

It's funny; sitting in a gun turret is a lonely task so you have plenty of time to think. You're scanning for threats, turning in multiple directions, but after a while other thoughts begin to creep into your head. I remember right about that time we were driving down the road, I had begun to get a little pissed because, while I had been in a few firefights, I hadn't been in anything I'd consider significant.

As we approached the footbridge, I looked left, then center, and then right, and at that exact moment, seeing those Humvees parked in what appeared to be some sort of static security presence, a quick thought passed through my head. "Jarion, you've missed the war. All you're going to be is a cop in Iraq."

No sooner had that thought crossed my mind than I heard rounds cracking over my head. Now, many people have probably heard rounds at a gun range, but they sound vastly different when they're coming at you. They say when a round sounds like an angry hornet or a pissed-off bee, and you feel the change in the air pressure—sort of an unnatural feeling—that's when you know it's real. The rounds were flying past my head and impacting against the side of the Humvee. Pissed-off hornets and weird air pressure were all around.

Later, we learned that seventeen enemy fighters had massed at the edge of an alley alongside Route Force and were about to attack one of the Strykers. They wanted to take down a large vehicle in hopes that the action would draw away more troops from the main attack. So, we just happened to drive through at the very moment of their attack, and my Humvee—which was between the insurgents and the Stryker—presented a target of opportunity. It was seventeen well-armed insurgents, who had the element of surprise, against four Green Berets and one Navy SEAL. You could say, "Wrong place, wrong time," but it'd describe the insurgents ... not us.

After hearing those first rounds, my initial thought was, "Oh, shit." I saw that the fire was coming from the right, so I quickly ducked behind the turret's armor panels and then heard the driver shout up to me, "Right side! Right side! Right side!" Then I felt the vehicle pull to the right.

I peeked over the turret's armor and saw the insurgents firing from the edge of the alley's entrance. So I spun the minigun in their direction with one hand, flipped off the safety, toggled on the power, hit the trigger for two thousand rounds per minute, and sprayed them with a few hundred rounds. With that one action, we gained instant fire superiority. Within a second the situation went from us taking dozens of rounds from enemy rifles to receiving nothing at all. Zero. Now they were scrambling.

The standard practice at that time of the war was for American troops to drive out of an ambush and never really engage. Americans weren't stopping and fighting back. The idea was to get away, regroup, and then come back ... maybe. The insurgents had mounted enough ambushes to know this tactic as well, and perhaps they were hoping to kill us and maybe the guys in our second vehicle, too, but they picked the wrong dudes to mess with that afternoon. Instead of driving out of the ambush, we went into a full counter-assault and took control of the situation.

As I fired the minigun, the driver maneuvered our Humvee into a power slide off of the main highway, down an embankment, onto the access road, and then started chasing the insurgents. They were now all running down a narrow alley towards three getaway cars. So they were already channelized ... it was like shooting fish in a barrel.

The first two guys I saw were carrying PKM machine guns and, I hate to say it, but they were overweight and obviously couldn't keep up with their buddies who were already down the alleyway. I cut those two guys in half with the minigun. The rest of them were all scrambling, shoving each other to get into their cars while shooting back at us. I pressed the trigger and started chewing up the first vehicle, and then the second, and then back to the first, all while the driver was tearing ass down the alley toward the targets. We were about seventy-five meters away from the first vehicle at this point.

One guy stepped out of the passenger seat of the second car, and I could tell he was shot in the leg and elbow, and his AK was still held in his dangling arm. I caught him with the minigun and he just came apart. Then I continued engaging insurgents all down the alleyway. I'm not claiming to be the baddest gunner on the planet or anything; it was just a target-rich environment.

One guy stepped out of a car with an RPG and shot it straight at our Humvee. Thankfully, he wasn't very experienced with the system and anticipated recoil as he fired. As I was firing the minigun at the vehicles I saw the grenade launch, hit the road, skip across the pavement, and felt it detonate on our engine block. That block took the brunt of the blast, which saved our lives, for sure.

So now our vehicle was on fire. The driver didn't lose control completely, but the brakes were gone. We slammed into the back of the first vehicle. All of those guys were dead by then. My earlier rounds had torn open the car's gas tank, though, so it caught fire from the flames coming from our Humvee. It was a mess, but I didn't let off the minigun. I was still spraying rounds into the second and third vehicles.

The next thing I heard was the driver shouting from below, "Hey, Jar. Get the hell out!" I yelled down, "Roger that!" but stayed on the gun for some reason. Then the driver threw open the door and rolled out through the flames. Turns out, his only injury that day was burning his beard! He managed to combat roll into what looked like the alleyway's gutter, raised his M4, and took out one of the fighters near the third car—the guy who shot the RPG, I think.

That's when one of those moments of clarity happened and everything seemed to slow down a bit. What do they say ... clichés become clichés because they really do happen? Well, it's like that— time slowed, and all of the sudden I could sense everything with great detail. The fire. The heat. I saw what the driver was doing. I saw which insurgents were still in the fight. I knew where to fire the minigun ... and I also knew that I only had a few seconds to get off of that truck. Hell, it was on fire and there were thousands of rounds of ammunition in the trunk. Then I saw another guy with an AK, and hit him full on with the minigun.

Finally, the slow-motion cliché gave way to the "little voice" cliché. It told me, "Hey, Jarion. You better get the hell off of this six-ton burning firecracker before it blows your balls off." That suddenly sounded like a great idea, so I climbed up and out of the hatch, jumped through the fire, and landed on my feet in the alleyway. I couldn't believe the jump didn't hurt anything. Then I raised my M4 and started laying down covering fire.

By then, probably just a few seconds into the whole fight, the second Humvee pulled up and the driver and I jumped inside. With all of the bad guys dead, our vehicle burning, and the alley about to turn into a virtual shooting gallery because of the ammunition inside its trunk, we took off. The driver looked over and asked, "Hey, Jar, are you hit?" I checked: arms, legs, and nuts. "Yeah, I'm good," I said.

Overall, it was a good day. I felt that we were completely focused and professional, almost like a wolf pack. We faced the threat, saw the target, and started to engage ... not going crazy or expending unnecessary energy, but handling it all in a very focused way. But I do feel bad about one thing. We had borrowed that Humvee from another ODA. It was listed as their responsibility on the Army's official property books, so those guys got stuck handling a bunch of paperwork because it was destroyed. So, I guess the moral of the story, along with carrying a minigun wherever you go, is if you're going to get blown up, make sure to do it in someone else's vehicle.

Several years later, I visited the Dillion Aero factory where that minigun was made. They were making a few modifications and needed some thoughts from the end-user perspective. They had heard about that incident in 2005, but had no idea that I was the guy using their gun that afternoon. So before leaving our headquarters, the driver and I signed a picture of the burned-down Humvee with the minigun. When I got there I gave it to the foreman and told him the story. He was so surprised that he stopped the production line and called everyone over to hear the story again. I thanked them all, sincerely, because their craftsmanship and expertise produced a weapon that saved our lives that day. Those people needed to know they weren't just working a job; they were working to produce things that protect and defend our nation ... and its warfighters.

The team's Humvee after the firefight in Baghdad,
May 2005 (*U.S. Army*)

After his actions that day, Sergeant Halbisen-Gibbs received the Army Commendation Medal for Valor. The narrative that accompanied the award noted that he "single-handedly killed 10 insurgents and destroyed two insurgent vehicles" and that his "courage under fire, with no regard for personal safety, protected his detachment members and defeated a determined enemy force." His fellow Green Berets quietly stopped calling him "Jar Jar." It was simply "Jar" from that point forward. His reputation within the Special Forces community was beginning to grow, and it would be cemented by a similar action two years later. Meanwhile, Sergeant Halbisen-Gibbs stayed in Iraq, learned from his brothers on the team, and kept "taking the fight to the enemy."

Two days after that ambush, Vice President Dick Cheney appeared on Larry King Live to discuss the conflict. After being asked about a timetable for departing Iraq, the vice president said, "We haven't set a deadline or date. It depends upon conditions." When asked if the war would drag on for ten more years, he said, "No … I think they're in the last throes, if you will, of the insurgency."

But the insurgency wasn't in its last throes. Al-Qaeda told its followers that Iraq was now the central battlefield in its global quest. The historical location wasn't lost on the jihadists, either. This wasn't simply another war for them; this was central to their centuries-long jihad. This was underscored in a July 9, 2005, letter to al-Zarqawi written by Osama bin Laden's deputy, Ayman al-Zawhairi, in which he wrote: "I want to be the first to congratulate you for what God had blessed you with in terms of fighting battles in the heart of the Islamic world, which was formally the field for major battles in Islam's history, and what is now the place for the greatest battle in this era..."

Meanwhile, back in the United States the debate about our involvement in Iraq waged on, although the discussion was being had among an increasingly smaller group of people who either had some level of influence on the direction of the war, or the small community of warfighters and their families who were actually fighting the battles. Most Americans simply went about their daily lives as if we weren't fighting a war—two wars, in fact.

"People are talking about bullshit," Chris Kyle said to his wife as they listened to the radio while driving home from the airport. The "American Sniper" had just returned after one of his deployments to Iraq and was angered to hear that his country wasn't focused on the war effort. "We're fighting for the country, and no one gives a shit."

Democrats in Congress were nearly united in their calls for a withdrawal, and many Republicans were joining their ranks daily. Public opinion polls also showed that support for the war was declining. In early October 2005, CBS News reported that "more than half of Americans—55 percent—think the U.S. should have stayed out of Iraq (the highest figure to date), while 41 percent think taking military action there was the right thing to do." That same month saw the opening statements in the trial of Saddam Hussein in Baghdad, and our nation mourned as the number of its warfighters killed in Iraq crossed the two thousand mark. There weren't any signs of it stopping, or even slowing down, in the foreseeable future.

Our warfighters were deep into what many were seeing as both a civil war in Iraq and a global battle against Islamic extremists. The American public was divided. Congress was in an uproar, and the White House and the Pentagon were grasping to see any substantial and lasting success. There was a real sense that we were officially adrift. "Our mission in Iraq is to win the war," read a line in the National Strategy for Victory in Iraq, released by the National Security Council in November 2005. The fact that such a statement was even needed reveals a great deal about where our national discussion was at that point.

The year ended like it began, with an election on December 15, 2005. This time it was to elect a parliament for a four-year term, and rather than boycotting the election, the Sunnis turned out in large numbers. Attacks were down, as well. It appeared, at least for the moment, that the nation was cautiously accepting democracy ... yet the insurgency against the American-led coalition raged on.

CHAPTER EIGHT

"THE PLACE HAD COMPLETELY FALLEN APART."

ERIC GERESSY
SOLDIER'S MEDAL (NOMINATED), *SAMARRA*

O ur overall strategy since the invasion was easy to explain, yet it proved incredibly difficult to implement. By mid-2005, a growing number of experts believed it was also fundamentally flawed.

"The objective is to get Iraq under control at a basic level, train up Iraqi security forces, turn over responsibility to the government, and leave," explained military historian Frederick Kagan in the documentary Losing Iraq. He was joined by Jack Keane, a retired general, who added, "When you look back on that and analyze it, it's a short war strategy. Nowhere in there is a plan to defeat the insurgency." Thomas Ricks, a writer for the Washington Post who covered the war, put it more bluntly: "You had war tourism—units based on big forward operating bases, FOBs, going out and doing patrols from Humvees, and then coming back to their base. If that's the way you're operating, you're not in the war. You're simply a war tourist."

Al-Zarqawi's network of foreign fighters was launching suicide bomb-ings weekly, hoping to spark a civil war between the country's Sunni and Shia populations. Crime was rampant, the Iraqi government was inept, and its military was mostly unwilling or unable to fight. Back home, newscasts were ending their shows by listing the names of the newly dead, and articles that questioned our continued involvement were written hourly. America had gotten more than it bargained for in Iraq, and *wanted to get out as soon as possible. Some wanted to leave only after victory, some wanted to leave soon but with honor, but most just wanted to leave. The strategy of training Iraqis, handing the fight to them, and then going home sounded good to nearly everyone's ears, but there were some who knew it wouldn't work.*

Our commander in Iraq at that time, General George Casey, believed the overall strategy was on track and would eventually succeed. Still, in the summer of 2005, a discreet "red team" of military and civilian experts was established in Baghdad to challenge the strategy and propose alternatives, according to the book The Endgame, by Michael Gordon and Bernard Trainor. What the team learned was troubling. As reported by Gordon and Trainor: "The effort to disrupt the insurgents' planning had not been decisive. The enemy was able to retain freedom of movement and conduct operations. Iraq's security forces ... lacked qualified leaders. The development of the police was a year behind the Army's estimates ... Time was running out."

Casey's command staff had "its own metrics for determining progress and many of them, it seemed, had to do with real estate," wrote Gordon and Trainor. "Success depended on how much battle space could be handed off to the Iraqi military ... and how quickly the United States could shrink its own network of bases."

Regardless of the National Security Council's pronouncement that our goal was "to win the war," it appeared that our goal was only to leave with honor, and even that was now in question. Gordon and Trainor explained that the red team's analysis "affirmed that Casey's strategy was badly off course and had almost no prospect of success" and that his "timetable for handing power over to the Iraqis simply set them up for failure."

Instead of making plans to withdraw American troops, the team of experts recommended what amounted to a "mini surge" of coalition forces into key areas to establish security first, and then work to develop the area politically and economically. This would be done in an "ink spot" fashion, they wrote, covering the most influential areas until the scattered spots could slowly melt together, eventually pushing out and isolating the enemy. It was a traditional counterinsurgency strategy, but it would take more time and more troops than Casey's train-and-leave strategy had called for.

It would be a hard sell, as Casey and others believed that the very existence of American troops in Iraq was why the insurgency survived. Once we left, the thinking went, the insurgents would leave or be driven from power.

After weeks of review, the red team prepared to present their findings and recommendations to the coalition's senior military and diplomatic leaders on August 23, 2005, at the U.S. ambassador's office in Baghdad. It didn't go well. "Casey took the floor and began to question the rationale for a red team assessment," wrote Gordon and Trainor. "The subject dominated the discussion, and the red team's PowerPoint slides were never presented."

Some believed that our military leaders had fallen in love with their plan and were unable to dramatically alter its direction. Our political leadership was too eager for a withdrawal to question the plan's assumptions, either, and the American public was impatient for the news to change. Casey's plan to leave Iraq remained. "An opportunity to change the strategy by making counterinsurgency the core of the military's approach had passed," wrote Gordon and Trainor. "It would take another sixteen months before Casey's strategy of handing off to the Iraqis was formally rejected and for the concepts in the ink spot approach to return in the guise of the surge."

It's important to note that the notion of counterinsurgency was being discussed and a few of its elements were being implemented in the field, although unevenly and often at odds with other parts of the strategy. For instance, Casey required that all incoming commanders attend a brief counterinsurgency course that he had established in Iraq. His overall plan also included bringing public services and employment into areas hostile to the new Iraqi government. But the larger parts of Casey's strategy that would keep our warfighters stationed on large bases away from the population, turn over the fight to an unprepared Iraqi force, and then leave as early as possible, cancelled out any benefits gained by trying a few counterinsurgency tactics.

Around the same time as the red team arrived at its recommendation—a counterinsurgency campaign with more troops—elsewhere in our military establishment others were coming to the same conclusion. Colonels leading brigades in the far reaches of Iraq, former commanders who cycled back to stateside positions and were now mulling over what they saw and learned on the battlefield, strategists in the Pentagon, advisors in the White House, and even a few writers and professors thought the same. They were mostly working independently, but within a year all would unite under a banner that would take a more evolved version of the red team's ideas from a PowerPoint briefing that was dismissed, into a game-changing strategy presented in the Oval Office.

Meanwhile, our warfighters were left to follow a failing strategy and watch as Iraq simply tore itself apart. Our troops couldn't haul a load of supplies down the road without the serious fear of being blown up by a roadside bomb, planted in plain sight of the people they had liberated only two years earlier. Suicide bombings at markets and mosques were a near daily occurrence, killing civilians—often women and children—indiscriminately. But it wasn't just the fighting that wore on our men and women in Iraq, it was the overall horrible condition of the nation and the people as a society. Decades of a brutal dictatorship and intermittent wars had left the people without a collective conscience. We wanted to help them, but to many of our warfighters who were sweating, bleeding, and dying in Iraq, the people didn't appear to want to help themselves. This made an already hard war even more difficult to fight. The Iraqi people were still in shock, it seemed, and in fairness they hadn't yet lived in the relative peace that would foster the sanity needed to govern a nation.

"Some days I thought we had broken into a mental institution. One of the old ones, from the nineteenth century, where people were dumped and forgotten," wrote Dexter Filkins in his book about the war. "It was like we had pried the doors off and found all these people clutching themselves and burying their heads in the corners and sitting in their own filth. It was useful to think of Iraq this way. It helped in your analysis. Murder and torture and sadism: it was part of Iraq. It was in the people's brains."

Eric Geressy, the soldier who stumbled upon Geraldo Rivera in the desert and then spent weeks guarding Baghdad's hospitals, was coming to understand that, as well. After his first deployment, the Staten Island native was now thirty-four years old and had been promoted to first sergeant of Charlie Company, 3rd Battalion, 187th Infantry Regiment, 101st Airborne

*Division. Geressy was still with his beloved Rakkasans, and they were back
in the fight. They deployed again to Iraq in the fall of 2005 where they
would see some of the heaviest fighting of the war, witness firsthand the
brutality of al-Zarqawi's network of Islamic terrorists, and learn that our
strategy of pulling back only emboldened our enemy.*

ERIC GERESSY: It was a horrible time to be over there. I was at the
tactical level and certainly not a strategic-level thinker, but I had been
around long enough to understand what was happening. Things on the
ground had gotten ten times worse from when I left in 2004. The place
had completely fallen apart. Al-Qaeda in Iraq was planting roadside
bombs left and right, and we were fighting the hell out of them everywhere. But it was painfully obvious to us guys on the ground that there
was no clear objective, no clear strategy, no clear mission, and no clear
guidance. The only thing that was clear was that the safety of the U.S.
troops was not the primary concern of some of the generals who were in
command in Iraq at that time. We were just supposed to drive up and
down the road until we got blown up or killed.

My perspective of some of the generals in Iraq in 2005–2006 was
that they thought we were there to give money to the Iraqis, to
rebuild their nation, to construct roads, clinics, and schools. It seemed
to me that their strategy was to throw money at everything and then
we'd be just fine. They were completely ignoring the enemy situation
that we faced in Samarra during that time period. It was like, "We
don't have to shoot people. We don't have to fight back. We can just
build all this stuff and the enemy will just go away."

That's such a dream. I think it might have something to do with
how our military sends its senior officers to institutions of higher
learning, like Harvard and the rest of them. They're taught what
other government agencies and non-governmental organizations do,
and how they can work with the military. This exposure and education is great, but these military leaders shouldn't forget that they are,
first and foremost, commanders of warfighters.

Some of them completely neglected the enemy situation on the ground. For one, the money we were handing out for projects was fueling the insurgency. They were using that money to buy weapons and explosives. It was so bad. The senior leadership, the generals in command in Iraq at that time, seem to have been focused on how much money we could give away, most of which ended up with the enemy. But our commander, a colonel, wasn't there to give money to the enemy. He was there to fight. The dysfunctional command relationships from the battalion up to Multi-National Corps Iraq put us all in a really bad situation, and in the end, the soldiers were the ones who paid for it.

For instance, at one point we received an order saying we couldn't detain a suspected insurgent based off of intelligence alone. We needed to actually see them shooting at us or planting a bomb somewhere, like how policemen work. But we were air assault infantry, not policemen, and this was a war. I read that order and then thought, "If intelligence drives operations, and we're not gathering any intelligence that we can operate from, then what the hell are we doing?" On top of that, our actions during every firefight were second-guessed. So, eventually the enemy was emboldened by those restrictions and things just got progressively worse.

Anyway, I had figured out during my first deployment that the only thing that really mattered was getting my guys home alive. That was it. As a company first sergeant, there was nothing greater that I could accomplish in Iraq than getting my boys home safely.

We were first sent to Baghdad in the fall of 2005 where our mission was to air assault into enemy-controlled areas and capture high-value targets. On a typical mission we'd load into the Black Hawk or Chinook helicopters in the middle of the night and fly to the target, sometimes setting down right next to the buildings that we would assault and other times landing a little bit away and then walking to the target locations. We'd do the raid and then bring the bad guy back to base. That's pretty much all we did from September until December of 2005, and it was a rough few months. I probably went to at least forty memorial ceremonies at Forward Operating Base Falcon for American troops that were killed in action during that time, and I will never forget those that were lost.

Eric Geressy, left, and Steven Delvaux of the 187th Infantry Regiment –
the Rakkasans (*E. Geressy*)

Eric Geressy, left, and Timothy Speece, in Samarra,
December 2005 (*E. Geressy*)

Sometime in December we were sent to Samarra, which is about 125 kilometers north of Baghdad. There was absolutely no functioning government there at that time. The governor, the mayor, the police chief, and anyone else having anything to do with governance had all run away or been killed.

Our mission was to secure a section of Main Supply Route Tampa, or MSR Tampa, which was our name for the interstate-type road that ran the length of Iraq from north to south. There were convoys of supplies and troops moving along that road constantly, and it was prime hunting ground for the insurgents. Snipers, ambushes, and roadside bombs were everywhere. It was our job to make sure that road was safe, or at least our part of it.

Our sector was about thirty kilometers long, which is a gigantic area for a company of 130–140 guys to patrol and secure. It was a dangerous section of road, too, especially since it had the intersections that went directly into Fallujah and Ramadi—not very friendly places. The unit we replaced had about thirty killed in action during their deployment, and that was just from their unit. I have no idea how many more were killed from other units who were simply driving down the road, trying to haul supplies from point A to point B. The unit that eventually replaced my company had about the same number of casualties, too. Thankfully, Charlie Company didn't lose a single soldier patrolling that route during the deployment. Either we were totally f'ed up or just really lucky ... or maybe it's because we chose to take the fight to the enemy.

We had everything you can imagine thrown at us—suicide bombers, IEDs, ambushes, mortars, and snipers. One day my guys called me on the radio saying they received mortar rounds impacting on their position along with receiving heavy machine gun fire. The soldiers immediately reacted to the enemy contact and killed an insurgent who was firing a DShK machine gun mounted on a motorcycle sidecar.

Now, a DShK is a very large Russian-made machine gun. It fires a round similar to our .50-caliber, and the overall weapon is about six feet in length. They usually mount those in the back of trucks or on large tripods. I couldn't imagine that they'd stick one on a motorcycle sidecar, so when I first heard the report I thought my guys didn't know what they were talking about. But then I saw it, and I was like, "You gotta be kidding me." The DShK was welded to the sidecar, and the guy used it ... at least until my guys killed him. It was the epitome of the "insurgent mobile."

The attacks along MSR Tampa were nonstop, so we figured we had to constantly patrol the road. We were always looking for any signs of wires along the road, which they'd use to detonate the bombs. This was before they really started using remote detonators. We had three outposts along our sector, all about squad-sized, enough for about eight men and two vehicles parked in a battle position overlooking the key terrain. We had one platoon manning those outposts on a rotation and then another platoon split into two groups who'd then patrol the road twenty-four hours a day. The platoon sergeant would take twelve guys with four vehicles and patrol for four hours. There'd be a four-hour break, and then the platoon leader, a lieutenant, would take another twelve guys and patrol for another four hours. We did that constantly. Often during the four-hour break I would take the headquarters platoon on a patrol, giving the rifle platoons a break. My headquarters guys were mainly the supply sergeants, clerks, medics, snipers, forward observers, and the mortar men.

MSR Tampa was mostly a two-lane road, but in some areas it would widen into four lanes—two lanes heading north and two heading south, separated by a median. We were near the Tigris River, too, so there were lots of bridges. It looked like any normal highway that you'd see in America. Heck, some of it was better than the roads you'd find back home in Staten Island, New York. Along the main road were dozens of old dirt roads leading to the farming communities in the area. East of MSR Tampa, toward the river, was farmland. It was mostly desert to the west, in the direction of Fallujah or Ramadi.

What were the people like?

Let me give you an example. We were patrolling the road on New Year's Eve. It was nighttime, probably about eight o'clock, and one of my guys saw what looked like a wire running from the side of the road and up into a nearby village. It looked like a picture-perfect IED setup. So we dismounted from our Humvees and maneuvered into the area.

We maneuvered to the cluster of buildings and started clearing some of the nearest structures, suspecting we would encounter the enemy. We had gone through three or four small buildings, going room to room, but found nothing. Then all of a sudden one of my guys called me on the radio: "First Sergeant, First Sergeant, you need to come see this." I told him, "Hey, we're still busy clearing a structure over here. Give me a minute." But he insisted, saying that I needed to get to his location quick. I was frustrated because he couldn't explain what I needed to see. He just kept saying, "I can't describe it. You have to see this for yourself." Damn, was he ever right.

I walked into the small building and was immediately struck by the smell. The stink was unbelievable. There was a guy sitting on the floor covered with a blanket. The poor guy was just sitting there, staring off blindly into space. His eyes were blank, like there was nothing there, like he had no soul. I reached down and pulled the blanket off of him. There was shit and piss everywhere, and worst of all, rats went scurrying off of him and into the shadows.

I remember thinking, "What the hell do we do?" We had no idea who he was, why he was there like that, or what to do next. We weren't trained for such a scenario, but like a leader must do in combat, you have to figure it out. I walked out of the building and saw a house about a hundred meters down the hill. I figured that'd be the best place to start asking questions. We walked down there and banged on the door. An Iraqi guy opened the door and said, "Oh, may I help you?" He spoke perfect English.

I said, "Yeah. Whose property is that up there," pointing up the hill to the building where we found the guy.

He said, "Oh, that's my property."

I said, "Really. What do you have up there?"

He said, "Just farm stock. Chickens and roosters mostly."

"Oh, just chickens and roosters?" I said. "How about a guy living like an animal? You got any of those up there?" One of my soldiers laughed. I guess he couldn't help it.

It was truly a bizarre situation. "That's just my brother," the Iraqi guy said. "Don't worry about him. He has no mind."

His brother? I was shocked. I didn't know what to say.

"Would you like to come in for some tea," the guy asked me. I remember thinking "God, no!" The last thing I wanted was to drink something from that guy. Who knows what would have been in it.

I said, "No, thank you. We're good." I was afraid to see what else we'd find in that guy's house. I didn't know what else to do, so we left. We had taken a couple of pictures for evidence because I knew nobody would believe what we found. We went back to the base and turned everything in to our battalion for follow-up. I don't know what they did with it, but it was bizarre. And just to give you an idea of what we were up against, those were probably the good guys in the area. The whole place was either part of the insurgency, supporting the insurgency, or treating their relatives like livestock. It was beyond belief.

Our regiment lost five guys killed in action during this time period, though. They had five guys killed in a Humvee when it rolled over a deeply buried IED. That's when the insurgents would dig up the entire road, the asphalt and everything, and bury a massive amount of explosives in the ground. They'd fill the hole, repave that section of road, and you couldn't tell anything was done afterward. What's worse is that these IEDs were probably being set by many of the same construction workers who we were paying to fix the roads in the first place. Our Humvees would drive over the spot ... and that'd be it.

What happened on January 9th?

I took the headquarters company out to patrol MSR Tampa that day. We drove for a while in our Humvees, and then dismounted and patrolled some of the unimproved roads that ran along the main route. Nobody had been out there in a long time, so I felt we needed to show a presence. It was early January so it was pretty cold. That was hard when we were patrolling at night, but it wasn't so bad during the daytime because we had all that gear on. The weather kept us cool, and the gear kept us warm. We walked for hours and hours that day; everyone was tired, and ready to get back to the base.

After we finished our dismounted patrol we got back into our Humvees and started driving in the northbound lane of the four-lane section of MSR Tampa. I was in the third vehicle in our four-vehicle convoy, and we kept about one hundred meters distance between each vehicle. The sun had just set and it was getting dark. We were about to transition into nighttime operations, which is when we would all put on our night-vision goggles, when the lead vehicle suddenly stopped. They got on the radio: "There's a box on the side of the road. It might be an IED."

By that point in the war, Iraqis knew to halt traffic in both directions when an American convoy stopped. A car speeding up to a parked Humvee would be considered a threat, of course, and might get shot. We also had signs hanging on the side of our vehicle, written in Arabic, warning Iraqi civilians to stay far away from our vehicles and deadly force would be used if they got too close. They eventually learned. The traffic stopped in all four lanes, northbound and southbound.

Now that we were stopped, and all traffic was stopped, too, our first move was to do what we call "5s and 25s." We first looked out of the window and ensured that there was nothing around us for five meters. Then, after everything was okay, we dismounted and checked up to twenty-five meters away from the vehicle, looking for anything like wires. Once that was done, the lead vehicle started to look at the suspected roadside bomb, the box, with their scopes and binoculars, trying to determine what it was.

While they were doing their inspection, I took out my handheld GPS and started to get the coordinates for our location in case we needed to call for support. I started working the device when all of a sudden I heard gunfire. I looked up and couldn't believe what I saw: a large black Chevy Suburban had pulled out of the southbound lane, jumped the median, and was heading straight for our lead Humvee as fast as it could go.

The gunner in the lead vehicle immediately opened fire with his M250 machine gun. It was certainly the right thing to have done. I think he killed the driver with the first burst of fire because the Suburban then swerved off-course and drove right between the first and second vehicles. Then the gunner in the second vehicle opened fire, and my vehicle's gunner opened fire, too, both guns striking the Suburban again. There was a massive explosion inside the vehicle, flames shot out everywhere, and it went flying off the road, *Dukes of Hazzard* style, before coming to a stop about thirty meters into a field.

It was still on fire, and there were secondary explosions, too. It wasn't the gas tank or anything like that. Something else was exploding in there. My initial thought was that it was loaded down with explosives, probably something large like an artillery shell that hadn't exploded yet. Maybe we shot the guy who held the detonator. The other explosions were probably grenades or large machine gun rounds.

I thought, "What the hell just happened?" I told my gunner to cease fire and tried to radio my guys in the first vehicle. I couldn't raise them, though. Then, to my utter amazement, I saw them leave their Humvee and run towards the burning and still-exploding Suburban. It was Sergeant First Class Ronald Newman and Staff Sergeant James Stephens.

I tried the handheld radio again. "Get the hell away from that thing!" I yelled. But our handheld radios came from the lowest bidder and did not seem to work when needed most. Newman and Stephens kept running towards the vehicle. I had no idea why.

I told my driver to use our vehicle's radio and tell headquarters what happened, and then I ran into the field, still yelling at Newman and Stephens to get away from the vehicle. I thought the main charge was about to explode, and that they were going to get hurt or killed. The secondary explosions were still going off, and the Suburban was still in flames. I yelled again, but as I got closer I saw what made Newman and Stevens run towards the vehicle: in the flames were the silhouettes of several children ... and they were on fire.

I cannot describe what that's like. Sickening. But what are you going to do, right? I ran towards the vehicle and joined Newman and

Stevens. The heat from the fire was intense, and shrapnel from the secondary explosions inside were whistling by our heads. We opened the door, reached in, and started pulling people out, kids first. There were three kids and four adults who were still alive. Two adults in the front were dead. The survivors were all screaming. They were shot, bleeding, burning and full of shrapnel. It was a nightmare.

Stephens, Newman, and I carried everyone up to the roadside and laid them down as gently as we could. They were in so much pain. I called to one of the Humvee drivers, a kid named Specialist David Robinson [*later killed in Afghanistan in 2010*] to pull up and turn his lights on the wounded so we could see them. Once he turned on those lights I knew we had placed our men at risk. We could see the wounded better, but the enemy could see us, too, plain as day, in the headlights. But there was nothing else we could do. We had to help these people.

I called to our medic, Doc Beau Weeks, who was the most junior medic in our company. I had assigned him to the headquarters platoon so we could assign the senior guys to the rifle platoons. This was his first combat deployment, and he was about to get some serious exposure to trauma treatment. Aside from the three kids, there was a woman, too. I think it was their mother. She was probably in her late twenties. She was shot multiple times and was bleeding pretty badly. There were three other adults, all shot, burned, and bleeding. They were all conscious and screaming.

Doc Weeks looked up at me. "What should we do?"

The wounded were in sort of a line. I said to the doc, "You work from the right side toward the middle and Stephens and I will work from the left side toward the middle. Try your best to do an initial triage and stop the bleeding."

Just as I said that, Stephens said, "Hey, there's someone else down there!" We shined a light into the field and, sure enough, there was another survivor trying to free himself from the burning Suburban. He was a bloody mess, too. We ran down there and carried him back to the road, placing him with the rest.

In all, we pulled three kids and five adults from the burning Suburban. I called the command post and gave a good sit-rep [*situation report*] and let them know I'd soon be rolling in with several casualties. Our Humvees weren't outfitted to carry anyone extra so we commandeered a farmer's truck from the southbound lane. He and his son were happy to help.

I turned my attention back to the wounded. Doc Weeks was working hard and, as we were moving towards the middle, I heard him praying over the wounded as he worked. I'll never forget that.

Apache helicopters had arrived to provide air cover, and we loaded everyone into the back of the farmer's truck and started hauling ass to the outpost. The guards saw us coming and opened the gate. We didn't even stop. We just rolled right in. We pulled up to the aid station and it seemed like the whole battalion was out there to help. It was a chaotic scene. We downloaded the wounded and took them into the station where the senior medics and the physician assistants went to work. One of the guys we pulled from the Suburban had his heart actually stop three times. And I don't know how in the world that mother survived. If it weren't for Doc Weeks, the medics, and the physician assistant at our aid station, they'd have all died, for sure.

I remember, after it was all over, sitting in the aid station next to one of the little girls we pulled from the wreckage. She had to only be five or six years old. She was shaken, of course, but aside from her hair being burned she was okay. I pulled off my helmet and everything else and just sat next to her for a little while.

We later learned that the driver and the other guy in the front—the two who had died in the initial burst of gunfire—were al-Qaeda. Those bastards had commandeered the vehicle, loaded it with explosives, and forced that family inside—three children, their mother, and five other adults, some pretty young, too. Maybe teenagers. Al-Qaeda knew that our soldiers, upon seeing a bunch of people in a car, especially women and children, would hesitate before firing on a charging vehicle. A moment's hesitation is all that's needed most of the time. But it had just gotten dark and our gunners didn't see the kids until after they fired and the Suburban went rolling by them.

As strange as it sounds, they actually saved the lives of every civilian in that vehicle by firing on them like that, killing the driver and probably the guy who would have detonated the car bomb. If our guys hadn't fired, the Suburban would have exploded, killing everyone inside of it and our soldiers in the first Humvee. And then, on top of that, they would have all died if Stephens and Newman hadn't seen the kids and went running after the vehicle to pull them out.

First Sergeant Geressy, Sergeant First Class Newman, and Staff Sergeant Stephens could have all justifiably stayed away from that burning vehicle. Most other men would have, especially since it was exploding with bombs and ammunition that was meant to kill them. Yet they rushed into danger, forgetting their own safety, to save the lives of innocent Iraqis who had been used as human shields in a cowardly attack.

Eric Geressy, right, and Ronald Newman after a firefight in Samarra, 2005. (*E. Geressy*)

Their company commander, Captain Daniel Hart, recommended each for the prestigious Soldier's Medal, which recognizes individuals whose heroic actions "involved personal hazard or danger and the voluntary risk of life under conditions not involving conflict with an armed enemy." But each soldier carries a far greater award—knowing that their selfless actions resulted in eight innocent lives being saved. Still, the kidnapping of the Iraqi family and subsequent suicide cemented Geressy's opinion of his enemy.

Using civilians, especially children, as cover was a sick tactic of al-Qaeda. We saw this same scenario play out again several times during our deployment. They would get entire families and use them as shields. Al-Qaeda was completely ruthless. That is what our soldiers were doing on a daily basis, what they were confronted with nearly every time they left the outpost. Day after day, on multiple deployments, for years.

Despite the coalition's desire to measure and report numbers that supported their strategy and made the case for the planned withdrawal, the trend since the summer of 2003 continued—Iraq was progressively going downhill. The month after Geressy and his fellow soldiers saved the lives of those innocent civilians, the situation managed to worsen. Writing in his book, Why We Lost, *retired Lieutenant General Daniel Bolger explained the moment when Iraq went over the cliff: "It's not often that you can pinpoint exactly when a massive undertaking fails. For the U.S. in Iraq, though, the moment was obvious. At 6:44 a.m. on Wednesday, February 22, 2006, two bombs detonated inside the al-Askari mosque in the city of Samarra."*

Al-Qaeda in Iraq chose the famous gold-domed mosque for its religious significance, and it was all part of al-Zarqawi's plan to ignite a civil war between the Sunni and Shia populations of Iraq. An article in the next morning's New York Times explained the significance of the attack: "The shrine is one of four major Shiite shrines in Iraq, and the site has special meaning because 2 of the 12 imams revered by mainstream Shiites are buried there: Ali al-Hadi, who died in A.D. 868 and his son, the 11th imam, Hassan al-Askari. Also, according to legend, the 12th Imam, Muhammad al-Mahdi, known as the 'Hidden Imam,' was at the site of the shrine before he disappeared."

Geressy and his Rakkasans were on a mission earlier that morning on the other side of town. "We saw the dome explode," Geressy said. "All hell broke loose after that, and the enemy turned it up big-time."

For months, the Iraqi government and its coalition advisors had endeavored to build a unified government formed of the nation's various factions. Turnout at the previous year's elections was high, and there was a great deal of hope that the democratic process could help ease old wounds and provide a peaceful, or at least less violent, way forward. In April, the Shiite majority in parliament selected White House-backed Nouri al-Maliki as their nominee for prime minister, positioning him as the leader of Iraq's first full-term government since the fall of the Hussein regime. Many of the political goals had been reached, but just as al-Qaeda wanted, the al-Askari bombing derailed all progress, and the old ways returned.

"Here was the real Iraq, the age-old Sunni-Shia divide scored in blood," Bolger wrote, adding that the mosque bombing was just the latest of many grievances. "It was always something, and it always ended the same way: beheadings, shootings, stabbings, bombings, death retail and wholesale, violence as language." Casey's strategy and all of its successful measures were "all gone, vaporized, swirling around in the crater that used to be the al-Askari golden dome."

In early June, a pair of U.S. Air Force F-16s sent two five-hundred-pound bombs sailing into the house where al-Zarqawi was sleeping. The head of al-Qaeda in Iraq was instantly killed … yet another quickly grew in its place and the beast continued to fight. Even though he was dead, al-Zarqawi had achieved the objective he had sought by ordering the al-Askari bombing. Iraq wasn't witnessing the last throes of the insurgency; it was in the middle of a civil war, and our warfighters were becoming little more than referees in a bloody melee.

"Most Americans in uniform once more asked that perennially urgent question. Who was the enemy?" Bolger wrote. "… the answer came back real ugly: Everyone. And they all had guns."

CHAPTER NINE

"IT WAS ALL VIOLENCE OF ACTION."

DIEM TAN VO

SILVER STAR, *AL-ANBAR PROVINCE*

I f the bombing of the al-Askari mosque in February of 2006 was
when Iraq fell off the proverbial cliff, then the announcement of the
Anbar Awakening the following September was when the nation
began its ascent from the abyss of civil war. The rope that helped them
climb up the slippery slope was a proven yet largely forgotten form of
warfare known as a counterinsurgency campaign. But to change directions
and discover the new way forward, our military strategists had to first
look backward to a book published in 1964 by a French military officer
and scholar named David Galula.

"A victory [in a counterinsurgency] is not the destruction in a given area of
the insurgent's forces and his political organization," wrote Galula, who
fought an insurgency in the Algerian War. "A victory is that, plus the permanent isolation of the insurgent from the population, isolation not enforced
upon the population, but maintained by, and with, the population."

Galula suggested a strategy that included, among other points, that a
counterinsurgent army had to mass a very large force and destroy or expel
the insurgents from a specific area, and then hold that area with enough
troops afterward to prevent the enemy from ever returning. Our problem:
The coalition had never had enough troops in Iraq to begin with, and our
goal to this point hadn't been to stay and hold an area, it'd been to tread
with a "light footprint" and leave as fast as possible. Furthermore, Galula

176

said that counterinsurgent forces should live with the population, gain their trust, and sever their ties with any insurgents, rather than redeploy back to large bases after the fight—which was exactly what American forces were doing.

Galula's book and other works on counterinsurgency had been making their way through military reading lists for decades and had found a small but energetic group of advocates. In the late 1970s, for instance, a young officer named David Petraeus read the book and became fascinated with its insights. By late 2005, there were many American officers in Iraq who wanted to employ a classic counterinsurgency campaign but were either blocked by superiors or hampered by resources. The strategy needed visionary field commanders willing to try it, but it also needed three actions that were not, at that time, politically viable: abandoning Casey's "light footprint" strategy, increasing the number of troops, and extending the mission's length for an unspecified period of time.

There was evidence a counterinsurgency approach could succeed. Aside from what Major General Petraeus did in Mosul with the 101st Airborne Division, Colonel H.R. McMaster successfully employed a classic counter-insurgency campaign in the northern city of Tal Afar in 2005 with his 3rd Armored Cavalry Regiment. His efforts were later described in the Washington Post as "distinct from the kind of war most U.S. commanders have spent decades preparing to fight," which was why so many other commanders weren't using them.

McMaster appeared on the Charlie Rose program two years later and explained the different ways of thinking at the time: "If you just take a look at it from a traditional ... military point of view that you have an enemy force, and you have to attack that enemy force and defeat it, it could lead you to sort of a 'raiding' approach to counterinsurgency ... Now it's an important capability in a counterinsurgency, to go after the enemy's network, gain visibility of it and attack it, but really you're not going to win because the key battleground in this war is a battleground of perception."

McMaster's success gained many counterinsurgency advocates in 2005 hoping for an overall change in Iraq, but none would come from Casey and the Pentagon. Tal Afar was a relatively small city of about 200,000, and critics noted that it was a far cry from the size and complexity of

Baghdad. The critique was sound. "It's a matter of scale," one officer in McMaster's unit told the Washington Post when asked if a similar counterinsurgency campaign could be launched in Baghdad. "You'd need a huge number of troops to replicate what we've done here." And that was one thing the White House wasn't going to approve ... at least not yet.

In the spring of 2006, the coalition pinned most of their hopes on the newly formed Iraqi government and its prime minister, Nouri al-Maliki. The White House wanted to give the new leader time to grow into the position and bring his nation under control, but all of the security trends were going in the wrong direction, and fast.

In the volatile western section of Iraq—the al-Anbar Province, which includes Fallujah and Ramadi—any local found helping the coalition or the Iraqi security forces was killed. The strategy to establish and counsel a new government, and train and advise a new set of security forces, was impossible to implement.

"The Americans tried to quell the insurgency in Ramadi with a combination of political maneuvers and the cooperation of tribal leaders ... but that plan has spectacularly fallen apart," read an article in a June 2006 edition of the Los Angeles Times. "The men who dared to ally themselves with the Americans ... quickly learned that the U.S. military couldn't protect them. Insurgents killed 70 of Ramadi's police recruits in January, and at least half a dozen high-profile tribal leaders have been assassinated since them ... Ramadi had become a town where anti-American guerrillas operate openly and city bureaucrats are afraid to acknowledge their job titles for fear of being killed."

Into that maelstrom rode the 1st Brigade of the 1st Armored Division, the "Ready First Combat Team," commanded by Colonel Sean MacFarland. He was familiar with counterinsurgency strategy and was willing to give it a try.

"In the summer of 2006, Ramadi by any measure was among the most dangerous cities in Iraq," wrote MacFarland and his operations officer at the time, Major Neil Smith, in the March–April 2008 edition of Military Review. Their brigade began implementing a counterinsurgency campaign, and certainly hoped it would work, but they wrote, "Few of us thought that our campaign would change the entire complexion of the war and push al-Qaeda to the brink of defeat in Iraq."

Ready First's plan was to attack al-Qaeda's safe havens, break up their ability to conduct operations, kill or capture as many of their fighters as possible, and gain the support of the people. "We intended to take the city and its environs back one neighborhood at a time by establishing combat outposts and developing a police force in the secured neighborhoods," MacFarland and Smith wrote. They also made sure to protect any tribal leaders who flipped to the coalition's side, even stationing armed patrols near their homes.

Along with fighting the enemy and protecting our allies, Ready First began to reestablish some of the critical public services that the people of Ramadi had done without, thanks to al-Qaeda's barbarism. One of their major efforts was to secure and improve Ramadi General Hospital, which had mainly been used to treat wounded insurgents rather than the local population. They also told local leaders about the brigade's different approach. "Instead of telling them that we would leave soon and they must assume responsibility for their own security, we told them that we would stay as long as necessary to defeat the terrorists," MacFarland and Smith wrote. "That was the message they had been waiting to hear. As long as they perceived us as mere interlopers, they dared not throw in their lot with ours."

The plan worked. Ready First chased the bad guys from the city, and moved in behind them to live with the people, protect their leaders, and establish badly needed—and greatly appreciated—public services. Because American soldiers were now living in the neighborhoods, the terrorists couldn't return to their nests ... so they stayed in the desert wilderness, and began to slowly fall apart. As the situation improved, more locals joined the security forces, as well, and more civilians were willing to take a greater, and more public, role in the running of the city. In late August a suicide bomber attacked an Iraqi police station, but in the aftermath the policemen didn't run away and refuse to return, like they used to after past attacks. MacFarland and Smith recalled that the Iraqis "remained at their posts, ran their tattered flag back up the flagpole, and even began to conduct patrols again the same day." That act of undaunted courage in that little police station signaled the tidal shift that came on September 9, 2006, when more than fifty local sheiks officially declared the "Anbar Awakening" underway. In a few weeks, nearly all of the tribes around

Ramadi also declared their support. This was a pivotal moment in the war, and al-Qaeda was finally on the run in Ramadi.

While the more irregular elements of the counterinsurgency strategy gained most of the attention—living with the population, protecting local leaders, etc.—one crucial element can't be overstated: the importance of simply killing the enemy. "You know how Ramadi was won?" asked Chris Kyle, whom the insurgents called "The Devil of Ramadi" for his prowess on the sniper rifle during the campaign. "We went in and killed all the bad people we could find ... The tribal leaders saw that we were bad-asses, and they'd better get their act together, work together, and stop accommodating the insurgents. Force moved that battle. We killed the bad guys and brought the leaders to the peace table. That is how the world works."

Kyle was right. Counterinsurgency is often misunderstood as being somehow a lesser form of fighting than the more muscular-sounding "conventional warfare." Some think it can be fought with fewer troops, less lethality, and perhaps a little less effort. But the experts say it actually requires more troops because you must own every square inch of the battlefield. It means you have to fight the enemy until he's completely beaten, without letting up. If an enemy fighter can be reconciled to peace—and there are always a few—then he may be allowed to live to help build that peace. If he cannot be reconciled, then he must be killed, and quickly.

Counterinsurgency also requires a great deal of risk, strength, patience, and even restraint in order to "win the population." An expert warfighter like Kyle knew that, too, and he often spoke about how he would take great care to ensure that all of his 150-plus kills were justified, although he noted that some of the "Rules of Engagement" he and other warfighters fought under were more about keeping commanders out of jail than actually killing the enemy.

And after all of that, adopting a counterinsurgency campaign will probably mean taking more casualties because your forces are dismounted and in the mix more than ever. They're not deep inside of their bases, or even inside of their armored vehicles. They're on the streets, with everyone else. Tanks and bombers aren't your greatest asset in a counterinsurgency campaign; the infantryman is. You're basically fighting to own every foot of ground, and then defending every inch of it for an unknown length of time.

Ready First's counterinsurgency campaign had worked in Ramadi, but elsewhere in Iraq it was still more of the same: car bombings, beheadings, sectarian violence, and full-scale civil war. Our leadership didn't see it that way, though. In a hearing in early August of 2006 before the Senate Armed Services Committee, the head of U.S. Central Command, General John Abizaid, tried to calm fears. "Despite the many challenges, progress does continue to be made in Iraq, and I am confident that there are still many more people in Iraq trying to hold that country together than there are trying to tear it apart," Abizaid said, adding that we could see additional troop reductions later in the year.

Sitting in his McLean, Virginia, home watching the Senate hearing on television was retired General Jack Keane. As recounted in the book The Gamble: General David Petraeus and the American Military Adventure in Iraq, 2006–2008, the retired general wasn't at all impressed by Abizaid's grasp of the situation nor Defense Secretary Donald Rumsfeld's dismissal of any need for additional troops or change in strategy. "My God, if we don't do something different, we're going over a cliff," Keane thought. The next day, he decided to do something about it, and began to speak with a small group of defense experts about the possibility of mounting a counterinsurgency campaign in Iraq in a last-ditch effort to win the war. Some believed it was already too late. Appearing on CBS's Face the Nation program on August 6, 2006, Senator Chuck Hagel—who would become the Secretary of Defense during the rise of the Islamic State of Iraq and Syria, or ISIS—called the war a "hopeless, winless situation." Thankfully Keane and a few others disagreed, and they began to plant the seeds for what would become known as "the surge."

Meanwhile, MacFarland's small-scale counterinsurgency strategy was making a big difference in Ramadi, but a battle plan is only as good as the warfighters who are executing the mission. Thankfully, Ready First was full of battle-tested officers, experienced NCOs, and dedicated soldiers. One of those warfighters executing the counterinsurgency campaign was 2nd Lieutenant Diem Tan Vo. His father and grandfather had been Vietnamese soldiers who fought alongside the Americans during the Vietnam War, but they were forced to flee after we left and the communists took control. Vo was born in a Malaysian refugee camp a few years later in 1980, and he came to America when he was only three months old. "I feel that this country gave me opportunities that I never would have had if they didn't take us in," he told me. "Military service is a down payment we pay for our children. So as a first-generation immigrant, this is my down payment to my homeland. It's a privilege to be an American."

DIEM TAN VO: I was doing my research rotation at the University of Chicago when we were attacked on September 11, 2001. Classes hadn't started yet that morning and I remember walking to the biggest building on campus. I turned on the television and realized, "Well, I guess I'm not getting my PhD ... Son of a ... the world just flipped!"

But you could have stayed and finished your doctorate, right?

And get left behind? Come on, man. This was the beginning of a whole new adventure. So after graduation, and when I was twenty-two years old, I enlisted in the U.S. Army. I'd always wanted to jump out of airplanes and be an infantryman, so I became an 11 Bravo, an 82nd Airborne paratrooper.

What was your college major?

Political science.

And that is a prerequisite for jumping out of airplanes, right?

I would imagine so. [*Laughter*] As you leave the plane, you're thinking, "Man, this is a bit of an existential crisis." The education allows you to examine it in its granularity and realize, "Hey, this might be a dumb call."

I was in Iraq immediately after the invasion in 2003. It was a lot of fun, actually, just the guys and me. Early on, after the fighting stopped, things weren't nearly as bad. I stayed an enlisted soldier for twenty-six months, made it to buck sergeant and then realized, "You know what? I'm done drinking cheap beer." So I got my commission and led an infantry platoon a few years later.

As a former sergeant, I learned from good and bad examples on how to be a proper officer. A lot of it is learning how to trust my guys, making sure they were trained well enough, and that they were making their own decisions. Having given command guidance, I preferred that my guys made their own decisions and then reported back. That's what sets the American military apart from the other militaries in the world. Aside from the Brits, the Australians, the Kiwis [*New Zealanders*], and the Canadians, most others don't have a solid non-commissioned officer corps like ours. Our guys are very well educated and very well trained. I have the utmost respect and trust in my men to get the job done right.

For instance, the American non-commissioned officer corps, uniquely among the first-world militaries, can have command authority in the field. There's no requirement for an officer, per se. I think it's uniquely American that the ranking guy on the ground, regardless if he's an enlisted non-commissioned officer or a commissioned officer, is empowered to make that call. As a matter of fact, he'll be in trouble if he doesn't make that call. It's his moral, ethical, and legal responsibility.

So, eventually I became a rifle platoon leader, a second lieutenant in Alpha Company, 1st Battalion, 36th Infantry Regiment, out of the First Armored Division—the Ready First Combat Team. We were heavy infantry, so we would deploy in fighting vehicles and then dismount foot soldiers. There were forty-two of us in my platoon.

I was about twenty-six years old when we deployed to Camp Hit in 2006. That's a city a little northwest of Ramadi, the provincial capital of al-Anbar. We were pretty self-sufficient. Quite a few times we were actually living off the land. During our patrols we would buy goats, sheep, watermelons, fish, bread, a lot of things like that because where we were was kind of hazardous to the resuppliers. We would slaughter the goat and cook it ourselves. It was good for camaraderie, and the meat was a good source of fresh protein for the guys. I remember one day, sitting around the fire, eating goat and thinking about the long line of unbroken infantry soldiers. I said to one of the guys, "You know, two thousand years ago a couple of Roman infantrymen were probably right here, doing and thinking the same thing; eating charred goat and thinking, 'Dude, this sucks. When are we going to get resupplied?'"

It was a good time. We were pretty successful with resourcing ourselves. One time we were in the middle of a firefight near the Euphrates River. One of the guys took out a can of Spam, put a chunk on a hook with some line, and then tossed it into the water. After we were done fighting we came back and he pulled up one of those weird-looking walking catfish that are in the Euphrates. The locals would eat those things. So we walked down to an Iraqi infantry stronghold and traded the fish for flatbread. A bunch of dudes stuck together in the middle of nowhere. We took care of ourselves.

Back then, in 2006, all of al-Anbar Province was inflamed with the insurgency. We'd begun changing our tactics around that time. We'd gone from keeping our elements inside large firm bases and moving amongst the locals. We started sending out scout elements to live throughout the city. It was funny, because our battalion developed the idea first before the 4th Infantry Division came along later and claimed it was their innovation. My commander and I were big history buffs and we were looking at some of the things that worked during the Vietnam War. We realized that the Iraqis only saw us as a bunch of invaders, a bunch of guys in heavy armor. We needed to integrate with the locals and actually become an in-place security element versus a bunch of dudes buttoned up on our firebases.

When we started doing that, the insurgents dug in and we started realizing there was natural selection at play. We had been knocking off all the wannabes, the "Ma and Pa" type of cottage-industry insurgents, and now we were finally going face-to-face with the hard core, whom many suspected were Chechens. Those guys can fight, I'll tell you. They knew their stuff. They were hardened veterans. We had a lot of respect for those guys as fighters.

Diem Tan Vo with an Iraqi soldier in the Anbar Province, 2006. (*D. Vo*)

Tell me about September 27, 2006.

We were at the company base that day, and my platoon was on quick reaction force duty. The sensors detected incoming mortar rounds and made a tentative fix on a point of origin. So my platoon rode out to try to interdict the insurgent mortar team before they ran away. We took two Bradleys, and about eight dismounted infantry and two non-commissioned officers. It took us about five minutes to get to the point of the mortar's origin.

It was launched from the middle of a built-up urban area, which made it extra hazardous because you never knew where they were shooting from. Urban warfare is the most technically demanding of all types of warfare. The streets were emptied at that point. People just vanished. They went inside their homes, buttoned up everything. There was no one in the streets.

We parked the Bradleys in a blocking position to prevent the insurgents from sending reinforcements into the area, and then we dismounted. We encountered sporadic sniper and small arms fire, and since we didn't know precisely where the mortar team was shooting from, we had to clear all the houses around there and find them.

We noticed this one building across the street. It had a big court-yard without overhanging power lines or anything to get in the way of mortar fire. So we figured that it was probably where they were launching the mortars. We were still under fire but we had to get in there. So one of my non-commissioned officers made the suggestion for us to "diamond up," and then run across the road and into the building.

Diamond up?

We put the non-commissioned officer, our communication guy and myself in the middle, like a diamond formation, and then put infantry around the outside of the diamond. That way, you have 360 degrees of security around the command element. Using that formation, we ran across the road, under fire, in a five-second rush and entered the building. There were about five dudes in there, and at that point it was all violence of action. We were on top of them before they had a chance to form into a kill team against us. We found traces of the mortars and then eventually the mortars them-selves.

A lot of times when they realized that they were outnumbered, they just tried to hide their weapons and other evidence. They'd scream, "We're innocent!" Then we'd find all their stuff hidden under the mattresses, and then it was like, "Not innocent anymore, are we?" A lot of times we'd capture them, bring them to the processing facility, they'd get detained for a few days, and then they'd be out again. They could count on a couple of dry nights in the detention facility, a few bucks for their trouble, and then get released. That's just the perspective of the guy on the ground. It was like a catch-and-release program, and kind of frustrating. But, it was what it was.

So after we cleared the building and secured the insurgents, we called in one of our Bradleys to load them up and take them back to the processing facility. This is when a lot of heavy firing started occurring. They started putting up machine gun positions down the street from us. It was some pretty effective fire, but using the armor of the Bradley as a shield, we were able to get all the detainees into the vehicles.

We were still taking fire, but we didn't want to indiscriminately shoot back because a lot of times the insurgents took up positions in people's homes, essentially using them as human shields. We would not fire at them unless we clearly identified the position that they were firing from.

So other than laying down the occasional sporadic burst of suppression fire from our Bradley's, we didn't engage. If we did, it would have resulted in massive amounts of civilian casualties because a Bradley's 25mm auto-cannon can punch right through a wall. You don't know what people's families are doing behind that wall. They might be cowering, kind of walled-in, thinking it's protecting them. They might have nothing to do with the insurgency. That's something we were keen on protecting.

Right around that time, my company commander, Captain Eric Stainbrook, showed up. He said he wanted to see the situation for himself. We were still under sporadic small arms fire but it wasn't effective, so I gave the captain a quick brief of what had happened. Then I went inside my Bradley to get on the radio and report the situation and how many people to expect in the detainee transfer.

That's when the ambush occurred. Back then, we didn't have things like drones circling overhead to give us constant coverage, so while we were clearing houses the insurgents would move along parallel streets to set up ambush positions. They started hammering away at us from three different sides. Captain Stainbrook got hit, the fire support officer, Second Lieutenant Bryan Jackson, got hit, and First Sergeant Sapp got hit. They were all bleeding out pretty badly, out in the open, too. They were in the middle of the street, which was the worst possible location to get shot because anyone who runs to help you is going to get shot, too.

I didn't really think. All I saw was three of my buddies down, and one of them was my boss! [*Laughs*] I remember thinking, "Screw it.

We're committed now. It's go time, dammit!" What else are you going to do? So I grabbed my medic and one of the riflemen and we ran out under fire, about 15–20 meters. It was instinctive. We had Americans down, and we weren't going to leave our buds out there. It does not matter what their ranks were. You do not leave your guys.

I had the Bradley start backing up to provide some semblance of support, because at least the Bradley could absorb rounds. We could not.

Did the insurgents see you?

Hell yeah. They started shooting at us. We had radios on, all of the officers and senior non-commissioned officers, and they could see the little antennae whipping behind you. They knew where to shoot. Like, "Hey, peg the guy with the antennae. He might be important."

The other platoon had ridden up at this time, so we had two Bradleys trying to close the distance between the casualties and the enemy to provide cover. All three of the casualties were on the ground bleeding out, but they were still returning fire. As we ran towards them we started waving and saying, "Don't shoot at us! Don't shoot at us!"

We got out there and started picking the guys up, trying to move them into the vehicles. I didn't really notice the rounds impacting all around me. I didn't really pay attention to that. It was just pure adrenaline, man. We picked the dudes up and threw them in the back of the vehicle. Looking back, I'm sure we could have been far gentler to our guys. I remember thinking to myself, "Dude, first sergeant is a moose! He's huge! I'm gonna need two more dudes to carry him outta here! If we survive this, I'm going to give him so much crap for being a fat ass." He was not a fat ass, really. He was just big, and with all of our body armor on, he weighed a ton.

After we got everyone safely into the Bradleys I did a last-minute check to make sure everything was picked up. You don't want them to get your crypto gear or your weapons. So, after a quick look I saw that some weapons and other sensitive equipment, like combat gear, was lying out in the street. So, me being a dumbass, I ran back out under fire to grab all that shit. After I picked it up and was coming back I got shot through the right tricep. It felt like a charley horse. Didn't think it was that big of a deal. I figure I just got winged, but apparently it cooked right through.

I dove into the sewage canal on the side of the street, and I was pissed. I was like, "Man, enemy marksmanship is so bad. They just got lucky." I was able to make eye contact with my guys. They identified where the insurgents were shooting from, swung the main turret on the Bradley around, and then blasted a hole in the side of their building. That solved that problem.

That was the third or fourth time I ran under fire that day. They just got lucky. But I got all the gear, though, so I still won. [*Laughs*] I had a notebook in my shoulder-area pocket, and the round went straight through the notebook, through my tricep and out of the other side. I still have that notebook, and I'll tell you, it's the coolest 'man trophy' ever.

I was lucky, though. I could have been shot up like the others. The other three guys were hit really badly. I think Eric Stainbrook actually died a couple of times and was resuscitated. Our first sergeant nearly lost a leg, and the fire support officer was hit multiple times, a few times in the ass, too. We gave him shit for that later, of course.

Everyone survived, and we all had funny stories later. We lucked out, man. A lot of guys weren't that lucky. That was that. Everyone got buttoned up and we rolled out. They brought in a tank platoon to clear that sector of the city later on. I stayed in Iraq for another four and a half months.

You didn't get sent home after such an injury?

Nope. I was sent to a Marine hospital near Haditha Dam. I was there for two weeks. I spent the first week getting fat on ice cream, the second week trying to break out. I wanted to get back to my guys. I tell you, Marines are good dudes, but they are some dirty bastards. One day at the Marine hospital I started scratching and went to the medics.

I said, "Hey doc, what is this? I got little bumps all over me."

The medic said, "Sir, you didn't hang out with Marines, did you?"

I said, "Yes."

The medic said, "Oh. Well, sir, then you've got fleas."

I got fleas. Can you believe that? I got on the SATCOM [*satellite communications*] and called home to my wife: "Hey, babe. So what happened was, not only did I get shot, but I got fleas. Can you send a care package of stuff to get rid of the fleas?" She was a little pissed. She was like, "What? Really? Our unborn twins almost lost a dad and now you have fleas? We didn't sign up for this crap." I said, "I know, babe, I'm so sorry. But no shit, I got fleas."

A few days later two of my guys from the platoon showed up at the hospital because they were hit by mortar rounds. So there we were, flea-bitten and picking shrapnel out of each other's skin like a bunch of chimpanzees ... you can't make this shit up.

Two weeks later my battalion came by to pick me up to bring me back to the line. It was kind of embarrassing because I had gained so much weight eating ice cream in the hospital that I had trouble putting on my body armor. But probably the most traumatic part of that whole affair was when I got back to my platoon, all of my stuff had gone missing. I said, "Hey, dudes. You know ... I'm not dead yet ... but you went ahead and split my gear. You don't split gear unless someone's dead, and I'm not dead." The guys all figured, "Hey, man, the lieutenant got shot. We got dibs on his gear!" I was like, "No, give me my shit back." I spent three days trying to find all of my gear.

It was a good time, man. As crappy as that deployment was, it was a good time. Those were some of the best guys I'd ever served with. I still keep in contact with them. I've always encouraged them to get out of the Army and do something more with their lives, especially the enlisted guys. I've been writing them letters of recommendation for college and making sure that they have the best possible start in civilian life, because they've gone through some pretty rough stuff.

At one point, my platoon had 60 percent casualties. I figure that it's about time to let someone else have a turn at this whole war thing. They all need to go home, go to college, and take care of their families.

I think what sets them apart from a lot of people is, they said "yes." I was sitting around with a couple of my colleagues who are former military a few days ago. To a tee, we all agreed that we asked ourselves before we joined, "If not me, then who?" We went, and it had nothing to do with politics, absolutely nothing. Politics has no bearing when you're out in the middle of the street bleeding out.

Some of the hardest stories to tell are the ones in the letters you have to write home for kids who didn't come back. You don't really want to tell a story unless it's a funny story, you know? That's why I always try to bring as much humor as possible in all the stories. There's always a funny silver lining. But the real stories are ... well ... we had a kid killed on Christmas Eve [*Private First Class Evan Bixler of Racine, Wisconsin, who died on December 24, 2006, in Hit, Iraq*]. No one wants to hear those stories.

Knowing what you know now, would you go again?

Absolutely, man. As crappy as it was, this was the project of our generation. I was not going to be left behind. I feel privileged to be part of history and honored to have served with some of the toughest men America has ever produced. Especially the infantry, man. We were far from politically correct. Everything was fair play. It was like a bunch of wolf pups together in a very confined environment. But those were invaluable experiences. It made me who I am today. Yeah, of course I'd do it again.

For his actions that day, 2nd Lieutenant Vo was recognized with the Silver Star, and the success of his Ready First Combat Team was being noticed back in Washington, D.C.

In early December, Jack Keane, the retired general who had been advocating for a counterinsurgency campaign since early August, gathered a group of a few defense experts and military planners at the American Enterprise Institute, a conservative think tank, for a weekend that changed the course of the war. The energetic and motivated group spent three days drafting a strategy that called for a wholesale change in the coalition's mission in Iraq and for additional forces to be deployed, and soon. At the forefront of everyone's minds were the lessons learned from McMaster's experience in Tal Afar, and MacFarland's ongoing counterinsurgency campaign in Ramadi.

Keane happened to have a meeting with President Bush the following Monday, and he took the team's recommendations into the White House. He told the president that we should implement a counterinsurgency strategy, increase the troops, and replace the commanders ... or we'd lose in Iraq. Even though Casey and many in the Pentagon continued to lobby for the "train and leave" strategy, and for additional troop reductions, Keane's ideas were taken very seriously in that Oval Office meeting. It was an event that was later called a "decisive" moment in the president's decision-making process.

A few days later, when Vice President Dick Cheney asked Keane if he'd like to return to active duty and lead the campaign, the old general declined. Keane did, however, offer the name of someone he thought was best equipped for the role: Lieutenant General David Petraeus, who, coincidentally, had just spent a year writing the Army's new doctrine on counterinsurgency operations. Petraeus was the right man, at the right time.

The year 2006 ended with Saddam Hussein being hung on December 30, and with the three thousandth American death coming on New Year's Eve. It was now clear to the White House that if we wanted to win the war, more warfighters must be sent to Iraq, and that even more would probably lose their lives before we could leave.

PART III

THE SURGE, 2007–2008

CHAPTER TEN

"I'M NOT GONNA WAIT."

CHRIS WAITERS, DISTINGUISHED SERVICE CROSS
JOSEPH MILLER, BRONZE STAR WITH VALOR
BAQUBAH

I*n early 2007, the war in Iraq was nearing its fourth year, surpassing the length of time American forces had fought in World War II. The nation grew war-weary as the cost of the conflict rose, both in blood and treasure. There were calls for a complete withdrawal, even from Congressional leaders who had been vocal proponents of the war years earlier. On January 10, 2007, the results from an ABC News/Washington Post survey that asked Americans if they thought "the war with Iraq was worth fighting, or not?" showed that only 40 percent said it was, while 58 percent said it wasn't. Later that evening, President George W. Bush addressed the nation from the White House announcing what the New York Times called "a major tactical shift." The president had decided to adopt Keane's recommendations to launch a counterinsurgency campaign, which would eventually become widely known as "the surge."*

Bush began his remarks by explaining how the violence in 2006 had "overwhelmed the political gains the Iraqis had made" in the 2005 elections. "Al-Qaeda terrorists and Sunni insurgents recognized the mortal danger that Iraq's elections posed for their cause, and they responded with outrageous acts of murder aimed at innocent Iraqis," Bush said, adding that Iraq's Shia population responded to the attacks by forming "death squads" and sending the country into a "vicious cycle of sectarian violence."

"It is clear that we need to change our strategy in Iraq," Bush said. He explained that the new strategy meant that Baghdad must be secured first, especially since 80 percent of the sectarian violence occurred within thirty miles of the capital. To do that, things had to change. Previous efforts had failed because of two reasons, he said. First, there weren't enough troops to hold areas that had been cleared of enemy forces. Second, there were too many restrictions on where our troops could go, and what they could do. "So I've committed more than twenty thousand additional troops to Iraq," he said. "The vast majority of them—five brigades—will be deployed to Baghdad."

Numbers changed, but so did tactics. "This time, we'll have the force levels we need to hold the areas that have been cleared," Bush said, hitting on a key part of counterinsurgency operations. American forces would also have the "green light" to enter neighborhoods that had been off-limits due to "political and sectarian interference."

Bush ended his speech by praising the "selfless men and women" of our Armed Forces who were serving in Iraq. "They have watched their comrades give their lives to ensure our liberty," he said. "We mourn the loss of every fallen American—and we owe it to them to build a future worthy of their sacrifice."

Five days later a roadside bomb in Mosul tore through the vehicle carrying Second Lieutenant Mark Jennings Daily of Irvine, California, killing him and three of his fellow soldiers. Before leaving on his deployment, Daily wrote an open letter to his family and friends, and anyone else, explaining the reasons he joined the Army during a time of war:

"Anyone who knew me before I joined knows that I am quite aware and at times sympathetic to the arguments against the war in Iraq. If you think the only way a person could bring themselves to volunteer for this war is through sheer desperation or blind obedience then consider me the exception (though there are countless like me) ... Consider that there are 19 year old soldiers from the Midwest who have never touched a college campus or a protest who have done more to uphold the universal legitimacy of representative government and individual rights by placing themselves between Iraqi voting lines and homicidal religious fanatics."

Daily's letter remains a timeless testament to the spirit and motivation of America's warfighters. But America wasn't sold on the idea of sending more troops. *A survey conducted in the days after his death showed that only 33 percent of the American public thought the surge would reduce violence in Iraq. There was also a widening gulf between the political parties, with 64 percent of Republicans saying they believed the surge was "somewhat likely to work," while 23 percent of independents and only 14 percent of Democrats shared that opinion.*

Bush was undeterred, and he followed up his announcement by accepting another of Keane's recommendations and appointing Lieutenant General David Petraeus to oversee the counterinsurgency campaign. Less than two weeks after the president's primetime speech, Petraeus sat before the Senate Armed Service Committee during a hearing on his nomination to replace Casey. He reminded the senators, "It is not just that there will be additional forces in Baghdad, it is what they will do and how they will do it that is important."

"None of this will be rapid," Petraeus said. "In fact, the way ahead will be neither quick nor easy, and there undoubtedly will be tough days. We face a determined, adaptable, barbaric enemy. He will try to wait us out. In fact, any such endeavor is a test of wills, and there are no guarantees." Petraeus then said that those who were serving in the military since the 9/11 attacks comprised the new "Greatest Generation."

Two members of that generation were Specialist Chris Waiters of Lacey, Washington, and Sergeant Joseph Miller, of Marrero, Louisiana, a small community just across the Mississippi River from New Orleans. They were both combat medics, and each came from a military family.

"Military service was sort of a family tradition," Miller told me. "Both my mother and father served in the U.S. Navy, and my brother, Shawn Miller, was a field artillery officer in the U.S. Army. I wanted to follow them into the service, so I enlisted in the army when I was nineteen years old."

Waiters said that his father, also a career soldier, told him that being a medic meant he'd probably spend most of his time driving around in an ambulance.

Miller and Waiters were in Iraq together when President Bush announced the change in strategy, which impacted both where their unit was sent, and how they fought once there. I first spoke with Waiters about

*April 5, 2007—a day nearly three months after the president's speech—
when he and Miller were serving with Alpha Company, 5th Battalion,
20th Infantry Regiment, in the town of Baqubah, Iraq.*

   "The surge" had begun.

CHRIS WAITERS: I was about twenty-six years old, and we were
deployed to Baqubah, Iraq. I can't remember the name of the base. I
told myself I'd never forget that name ... but now I can't remember.
The base itself wasn't bad. Baqubah was pretty bad, though. During
the surge a majority of the insurgents fled from Baghdad, because we
were surging with thousands of additional troops there. So the
insurgents went to Baqubah and set up shop there. The unit that was
up there before, well, they couldn't handle it anymore. They were
taking too many casualties.

   I was serving in an infantry brigade, but it was all Strykers, so we
weren't considered light infantry or armored. We were in the middle.
A Stryker was a new vehicle, and it was a new concept in the Army
at that time. Almost like a mini tank. So they called my battalion up
to Baqubah, about five hundred soldiers with our Strykers, to hold it
down before the rest of the brigade joined us.

   We had our own medical Stryker. Joe Miller, who was the senior
medic, and I would switch who would go out with the company XO
[*executive officer*]. We were doubled up, pretty much, so every other
mission we would switch. One day I would be on the ground running
around with the guys and Joe would be in our Stryker, and the next
day he would be outside and I would be in our Stryker.

What do you remember about April 5?

We were on a clearance mission in Baqubah that day. I remember
everything, every little detail. I think about it every day. It was quiet.
Not a lot going on. We were clearing: knocking on doors, checking
for weapons and insurgents. The company was clearing by foot
because of the deeply buried IEDs. They were planting bombs in the

roads, to defeat our Strykers from underneath. So we couldn't just be driving down the road. There were these Bradley units from Fort Hood and their main objective was just to keep the roads open and cordon off anything from the outside. We broke down a wall and stayed on the soccer field. So we set up shop there, my vehicle and then the company XO's vehicle. I was there so they could bring casualties back to me.

I want to say it was like 7:30 in the morning. It was a peaceful day, really. It was probably about eighty-five degrees. Sunny. Not a cloud in the sky.

I had pulled a guard shift the night before, sitting in the hatch, so I had been awake for probably twelve or thirteen hours, and I felt paranoid. I switched spots with Joe and laid my head down on the medical Stryker's stretcher for a quick second, and then I heard this massive eruption. Boom! The next thing I heard on the radio was, "Hey, Chris, let's move!"

We just started moving, fast, and then I heard over the radio, "We've got a tank on fire! There are guys burning up over here! We're in contact!"

We scurried around the corner, I want to say a half a mile, and all of a sudden there were people everywhere. There was a market over there; a safe area, sort of. It was a road with two-to-three-storied buildings on each side for about a mile long. We called it Market Street and at the bottom it was just shops, so you had everything hanging out and people were selling stuff. So there were probably five hundred or six hundred people that were in the road, just screaming and running. We stopped for a second. I faced my vehicle east. The XO faced his west, and as soon as I did, I saw two insurgents running down toward my vehicle with AK-47s. One had on a ball cap and jeans. The other one just had a flannel shirt, some sandals and some khaki-looking pants, but they both had AK-47s in their hands and they were coming towards my vehicle. I was still in the Stryker, sitting in the top hatch, so I immediately raised my weapon, an M-4 rifle, and killed both of them with about six or seven shots.

It was pure chaos. Joe was on the radio trying to formulate a plan. About 150 meters down the road, the whole street was on fire. There was this Bradley in the middle of the road, just burning and people running around it. I could smell it. I told my boy I was leaving. He told me to wait, but I said, "Joe, I'm not gonna wait." So I dropped the ramp, and the last thing I heard from Joe was, "If you go out there you're going to die!" I just took off down the road.

Fifty meters down the road there was a barrier—a small concrete barrier, the kind that goes up to about your hip. I ducked behind the barrier and that was when the insurgents popped up out of the windows, off rooftops, and from around the corners. They started shooting at me. There were seventeen total.

You counted them?

Yeah. There were seventeen, all around me, and all by the Bradley. That was their kill zone, from all angles. So, what I did was this: I killed three of them that were standing there. One was standing by a market where meat was hanging up for sale. He was ducking behind the meat stand, trying to blend in, but he was shooting at me at the same time. I killed him.

I saw two guys above him in the window that were shooting. I shot at those guys; they never shot back. I killed them, too. Then I immediately got up, reloaded and ran farther down the road until I got to a car that was about sixty meters from the burning tank. I ducked behind the car real quick, and there were about seven more guys on the left-hand side that were shooting. One guy had an RPG. At this point I was just trying to figure out what I was going to do next because, well, the moment of truth had arrived. I finally realized what was going on ... because I wasn't thinking at all after that first bullet cracked, but then, down the road, I was like ... "Man, I'm going to die."

Another truck came around with a gunner on the back and engaged me with a machine gun. It was a Toyota flatbed truck. We called them "bongo" trucks. He drove through the smoke of the Bradley, through the fire. There was a guy standing on top with a mounted machine gun, and he fired directly at me. Before I even got a chance to shoot him, my XO's gunner with a .50-cal destroyed that truck completely. Then he used their mounted weapon system to help me out the rest of the way down the street.

So I got up and started running again. When I ran by the bongo truck, it was still being engaged with .50-cal bullets. Both guys were shredded. They were gone. I raised my weapon and shot three more fighters who were behind a truck, over in a little deli area where watermelons were being sold. So, everything was hanging out for sale in the market street, there was a burning tank in the middle of the road, and there was me, running down the street, and people were scattered everywhere. Just running for their lives.

The tank was a complete ball of fire at this point. The whole back was engulfed because that's where the fuel cells are. The only thing that was not on fire or had flames coming out was the gunner's hatch and driver's hatch. Everything else was fully engulfed. I was about to get on the tank, but then another guy came out of an alley about twenty-five meters away. He throws a grenade, but it never blows up. So, I raised my gun and shot him, too.

After you had reached the tank, what were you thinking you'd do next?

I wasn't thinking. I just knew I had to get onto the Bradley. I took off my gear and my body armor because it was too heavy. Each plate weighs about twenty-five pounds, plus full ammo capacity. I left my weapon, too. The only vehicle that size that I was familiar with climbing on was a Stryker. I didn't have any idea about how to climb on top of a Bradley. I can't remember, but I just jumped on top somehow.

There were rounds pinging all around me on the side of the Bradley, and I was looking for an opening so I could get the guys out. Smoke was everywhere and I could hardly see. Then, by the grace of God, I saw a hand reach out of the hatch. I grabbed it and pulled him up. I looked over and saw another hand. I shouted, "Get up! Let's go!" I got them off the Bradley and put my gear back on real quick. They were just charred—charcoally, and had difficulty breathing. You could see the redness in their eyes, tearing up. They were exhausted.

I got them down on the ground. I put my vest on real quick and grabbed my gun. More insurgents were shooting at us, so I shot back to just suppress them. At that time, my unit had stopped their mission and had come down to help me. So I had snipers on the roof down the road. I had a whole platoon moving on my section at that time and now the insurgents based their fight on them. So I wasn't a target anymore.

A Bradley that had come to reinforce the fight pulled up next to the burning Bradley, and for safety I stuck both of those wounded guys inside its cabin so that small arms fire from the insurgents wouldn't affect them. We drove them up the road to my medical Stryker so they could get treatment. I got them into my Stryker and one of the guys looked at me as I was putting oxygen on him and said, "I've got a buddy in the back." I looked down and said, "I'll be right back."

That's what you told him?

Yes. "I'll be right back." Then I took off back down the road again to try to figure out how I was going to get back on the burning Bradley. It was still engulfed. They told me that the Iraqis were bringing a fire truck but I couldn't wait for that. I came around the back of the Bradley and kicked it open. Flames shot out. I climbed into the back, and in the back left corner, there was maybe a three- or four-foot area that was not on fire, but it was fully engulfed in smoke. I was wearing fire-resistant gloves, but it was still hot. I just sat still for a second. I remember thinking: *hot, bad situation.* I started trying to look around, trying to hear something, trying to hear somebody talk to me. But all I heard was crackling sounds from all of the plastic and metal in the tank melting.

It smelled like gas and smoke, too. I got overwhelmed, so I went back out, probably just for fifteen seconds to regain my breath, and then went right back in. Same thing as before. I felt around the interior and tried to hear for someone's voice. Couldn't see nothing but smoke and fire. Couldn't hear nothing but crackling. Then my boots caught on fire. I got out of the tank, put the fire out, and then went back in again. *[Waiters paused for a moment, and then his voice cracked as he continued.]*

The next time in the tank I finally found an arm and grabbed it ... but his whole arm came off like a cooked turkey leg. Like meat. His whole arm just peeled off. I went back out, caught my breath, and came back in, and then I grabbed his other arm and pulled him forward. But then ammunition in the back started cooking off and it blew his head off in front of me. I went back out again and ran back to my truck to get a body bag. My uniform was charred. I was black from all of the smoke. My boots were melted.

By then the shooting stopped. The people were gone because my whole unit was on the scene, probably 70–90 soldiers on one block. They were kicking in doors and going room-to-room in all of those buildings to try to find the insurgents.

The ammo in the Bradley was still cooking off. There were holes going through the Bradley's armor. I went back inside, and while I was in there, probably about forty rounds went off, and these were big rounds. I finally secured his body and carried him out of the Bradley. I had two medics on the ground at that time, so I gave the scene to them. The fire truck was on its way and I went back into my Stryker with Joe Miller. Gave my XO a sitrep [*situation report*]: "Two inhalation burns and one KIA. We gotta roll!" I called in the bird and ten minutes later the helicopter arrived. It took the wounded to another FOB [*forward operating base*] in our area.

Then, after it was all over, I just sat there. It was really hard to breathe. My buddies were like, "What's wrong, Chris? Are you okay?" My chest and back hurt, and I thought I had actually broken my back. Come to find out I'd been shot twice in the front and twice in the back but my bulletproof plates stopped the bullet. I was in pain for a little bit.

Do you have any idea what would have happened to those two guys in the Bradley?

They would have been dead. Those insurgents were waiting for the casualties to come out so they could execute them. That was a classic ambush. Some smart people orchestrated it. They weren't stupid insurgents like a lot of people said. Those guys were not stupid. They knew exactly what they were doing by the minute. That was very well planned.

What compelled you to run down the road that morning?

I don't know. I just saw hurt people. I'm a medic and that's my job. I didn't start thinking until I was pinned down and then I was like, "Holy crap, what am I doing? But I can't turn back now. I'm already here. So let's just continue the mission and do my best." But every day I always wonder: what if I ran faster, or what if I wouldn't have ducked, if I did this or that, or if I wouldn't have shot back? Every day I think about those things. Every day. It drives me crazy, just every day always thinking about that moment … what I could've done to change the outcome to something better. If I would've gotten there a minute faster, or even a couple of seconds faster? If I would've went to the back of the tank first instead of jumping on the front? Maybe I could have saved all three. It haunts me to this day.

What did you think when your commander said you were getting the Distinguished Service Cross?

I didn't want it. I didn't care. But it wasn't up to me. I didn't even know what it was at the time. A lot of people look at me and say, "It's a Distinguished Service Cross!" But you're actually probably the third person that knows the whole story. I don't really tell it a lot. I don't think it is anything to brag about, you know what I mean? People just say I'm humble, but I thank God that I am alive. And I always wonder *why* I am alive. Was the purpose of me staying alive to save those two?

Chris Waiters climbing atop the burning Bradley Fighting Vehicle in
Baqubah, April 2007 (*U.S. Army*)

The Distinguished Service Cross is presented to Chris Waiters, October
2008 (*U.S. Army*)

Chris Waiters, Joseph Miller, and their fellow soldiers were still fighting hard, but the outcome of the surge remained unclear. Many back home in the United States had already given up. Two weeks after Waiters killed several insurgents and rescued two soldiers from that burning Bradley, Senate Majority Leader Harry Reid held a news conference where he infamously said, "This war is lost and the surge is not accomplishing anything." Rasmussen Reports published the results of a survey in late April that showed 57 percent of Americans favored an immediate troop withdrawal, as well. The shifting public opinion harkened back to a bit of advice British Prime Minister Margaret Thatcher gave to President George H. W. Bush on the eve of the first war with Iraq in 1990: "This is no time to go wobbly, George." The political establishment and perhaps even most of America had indeed gone "wobbly," but our warfighters weren't about to quit, and the fight was far from finished.

Joseph Miller and I discussed May 10, 2007, just a month after that harrowing ambush in Baqubah where Waiters saved those two soldiers. Miller was with his unit on Combat Outpost Gabe, waiting for the next call. "If you are at FOB Gabe," wrote one New York Times reporter, "you are used to swarms of flies and mortars hitting the base daily. Like every Realtor says: It's location, location, location."

JOSEPH MILLER: Chris and I had split up that day because he had to cover down for a medic who was in another company. We were at the outpost and got word that Second Platoon was in a TIC, a troops-in-contact, situation and had been hit by an IED. I was part of the headquarters element that responded, and we rolled out of Gabe in a four-vehicle convoy to back up the platoon that was under fire.

Our commander and several riflemen were in the lead Stryker. A couple of mechanics were in the second vehicle, which was a wrecker in case we needed to haul a broken vehicle from the fight. My medical evacuation Stryker was third in the convoy, and riding with me was one of our snipers, Specialist Caesar Hernandez. The fourth Stryker was our executive officer's vehicle.

Within a few minutes of leaving FOB Gabe we entered the Old Baqubah part of town, an area with a lot of enemy activity. As we got near the other platoon we stopped because some of the locals had told us we were driving down a street that had IEDs buried under the road. I could hear the back-and-forth on the radio between our commander and some other guys talking about what to do.

After a moment, the commander decided to proceed. I'm not sure, but I don't think they even started rolling before the enemy detonated the IED. It was buried under the road, right under the front of the Stryker. The blast was massive. Smoke, dirt and debris shot into the air. It looked like a giant whale had come out of the ground. The Stryker, which is a very heavy vehicle, was thrown onto its side and rolled over, finally landing on its roof. It was blown apart in several areas; I could see smoke and fire coming from some parts of it.

Initially, being that close to that massive blast, I was in shock for a moment, but then started to quickly think, "What do I need to get ready for? What do I need to do?" My training kicked in.

The first thing I did was take my Bose headset off and then I yelled to the driver, Specialist Anthony Davidson, "Drop ramp! Drop ramp!" I ran down the ramp with my M4 rifle and my sidearm, a 9mm, and ducked behind a concrete barrier and tried to see what was happening along the street.

The scene was a classic ambush. Right after the insurgents blew up the first vehicle they started opening up with small arms fire from nearby windows and roofs. So I started firing back at the places where I saw muzzle flashes in the windows and along the rooftops. I'm not sure if I got anyone, but it at least suppressed them so we could move to the wounded.

I looked back and saw that my boy Caesar, who was the sniper, didn't have his rifle. He was running from our Stryker carrying two litters for the wounded. I remember thinking, "This isn't good ... the sniper has the litters and the medic has the rifle."

Then I ran past the second vehicle, the one with the mechanics inside, and made my way to the left side of the first Stryker, which was still burning. I knew there were about seven people inside who we had to account for, and most had already crawled out through hatches. They had recovered a little and were returning fire. The only two who were still inside were our commander, Captain Hubert Parsons, and the driver, Sergeant Jason Vaughn, who was a good friend of mine.

I could hear Captain Parsons yelling, so I climbed in the Stryker from its rear opening and started making my way towards his voice. The cabin was full of smoke and I could hear rounds pinging against the side of the vehicle, both from insurgent fire and some rounds inside the truck that were getting hot and exploding. Someone told me that one of our anti-tank rockets cooked off, but I don't remember that.

After a few seconds I found Captain Parsons. He was lying on the inside roof of the vehicle as it sat upside down. He was trapped, and was pretty banged up, with a compound fracture in one of his legs. He couldn't have gotten out of there himself, but I couldn't drag him out by myself because of the way the Stryker was mangled during the explosion. So I unhooked him and said that he'd have to help me by doing a sort of controlled backward crawl out of the rear hatch. By then the cabin was totally engulfed with smoke and fire, and more rounds were pinging around inside. We eventually made it into the clear and we put the captain on a litter.

I turned and looked back inside the Stryker. It was completely on fire, especially in the front where Jason Vaughn was located.

Major Timothy Price, who was the company's executive officer, would later write a memorandum capturing Miller's actions for the official record. "... Miller, with no regard for his personal safety, managed to evacuate and stabilize both the company commander and his interpreter, under fire, and while rounds inside the vehicle exploded as the Stryker burned," Price wrote. "He made multiple trips in and out of the kill zone, under steady enemy fire, to recover his comrades." He added that, during their fifteen-month deployment, Miller provided combat medical care for upwards of one hundred wounded soldiers and local Iraqis, and "continually shouldered this burden, always putting the needs of others ahead of his own well-being."

Miller was awarded the Bronze Star with Valor for his actions during the attack, but it's not those he saved that Miller remembers the most. It's those he couldn't.

We lost Jason Vaughn that day, and it hit me really hard. The other guys in his vehicle told me that he died instantly when the IED detonated … it exploded right under his driver's hatch.

I turned back to where we had taken the captain. I was pretty exhausted, and the rush of adrenaline mixed with all that smoke nearly made me sick. I pitched my rifle to one of the other guys and started working on the wounded. Then we grabbed the captain's litter, carried him to my medical evacuation Stryker, and then took him back to FOB Gabe.

I was frustrated … and really sad … that I couldn't immediately get to Jason and pull his body from the vehicle, but the fire, the wreckage, and the ambush were all too bad at the time. He wasn't recovered until we had secured the area.

Would you run into a burning vehicle again?

Yeah, of course. That's what we expected of every soldier in the outfit. If someone needed you, you responded. You laid down fire and moved to your fallen, and then got them out. That was instilled into the hearts of every soldier. That's what we do, and that's the way it has always been.

So, yeah, I'd do it again. It was my job, and they were my family.

If you could go back in time and talk to a young Joseph Miller before you enlisted in New Orleans, what would you say to him?

I'd tell him to do it again, no question about it. Looking back, I needed the Army more than the Army needed me. It reinforced the values that I was taught growing up, and it gave me a purpose, a sense of camaraderie, and it formed me into the man I am today. Some of the finest individuals I've ever met were in the Army, too. They're my family, and we continue to take care of each other.

We always will.

Joseph Miller, Basic Training, Fort Benning, Georgia, May 2001 (*J. Miller*)

CHAPTER ELEVEN

"WE'RE BLACK ON AMMO!"

JEREMIAH CHURCH
SILVER STAR, *BAQUBAH*

B y mid-June 2007 all of the additional "surge" troops, more than
twenty-eight thousand soldiers and Marines, had arrived in Iraq to
execute our new counterinsurgency campaign. Troops moved into
areas previously held by al-Qaeda and other insurgent forces to kill them or
run them off, and then armed some of the local tribes who pledged to fight the
terrorists. Our forces began moving into the neighborhoods, establishing small
outposts so that the local population would always see their presence, and so
that the insurgents couldn't return without a fight. It was a surge in terms of
the number of troops and their involvement in the fight, and also a surge of
new ideas that initially led to greater casualties. Still, General Petraeus and
his civilian counterpart, Ambassador Ryan Crocker, were confident that the
counterinsurgency campaign would work.

Back home, the American public remained skeptical. In a poll taken in
early July, only 19 percent considered the troop surge a success while 43
percent thought it was a failure. Congress, with both chambers controlled
by Democrats, was calling for a withdrawal, and many in the military's
establishment were quietly working behind the scenes to cut off the flow of
troops and return to the old "train and leave" strategy once it was clear
that Petraeus had failed.

By early July the White House had gone to full-scale political war with
Congress over the issue. Led by Speaker of the House Nancy Pelosi, a Democrat
from the San Francisco area, the House voted on July 12, 2007, in favor of

requiring a withdrawal of most troops from Iraq by the following April. The day saw dueling press conferences and speeches by leaders of both parties. "Sometimes the decisions you make and the consequences don't enable you to be loved," the president was quoted as saying by the New York Times. He said he'd stick to his strategy, and accused Congress of overstepping its bounds by attempting to direct military forces. Instead, the president asked Congress to reserve judgment on the surge until Petraeus and Crocker delivered their assessment in early September, setting the stage for one of the most anticipated, and consequential, Congressional hearings in decades.

But wait they would not. "It is deeply troubling that nearly six years after 9/11, al-Qaeda maintains a safe haven, an intact leadership and the capability to plan further attacks," then-Senator Barack Obama said in a statement issued on July 19. "It is time to act to correct those mistakes, and the first step is to get out of Iraq, because you can't win a war when you're on the wrong battlefield." The sentiment was echoed in a speech on the Senate floor by then-Senator Hillary Clinton, who at that point called for a complete withdrawal: "Our involvement in Iraq continues to erode our position ... It has damaged our alliances, and it has limited our ability to respond to real threats."

Meanwhile, things were indeed changing in Iraq, even though the news failed to reach the American public. Our politicians and journalists seemed stuck inside the narrative that had dominated the previous four years. But a powerful op-ed published in the New York Times in late July signaled a turning point in that narrative. Two highly respected scholars at the Brookings Institution who had been influential critics of the war— Michael E. O'Hanlon and Kenneth M. Pollack—wrote the essay, titled "A War We Just Might Win," and it spread like wildfire.

"Viewed from Iraq ... the political debate in Washington is surreal," O'Hanlon and Pollack wrote. "The Bush administration has over four years lost essentially all credibility. Yet now the administration's critics, in part as a result, seem unaware of the significant changes taking place. Here is the most important thing Americans need to understand: We are finally getting somewhere in Iraq ... we were surprised by the gains we saw and the potential to produce not necessarily 'victory' but a sustainable stability that both we and the Iraqis could live with."

The op-ed went on to describe how during previous visits to Iraq the authors found our troops "angry and frustrated" and suffering from low morale because they were "using the wrong tactics and were risking their lives in pursuit of an approach that could not work." That had changed, they observed: "Morale is high" among the troops, they believed in their new commanding general, they were "confident in his strategy," and finally felt they had enough troops to get the job done.

O'Hanlon and Pollock also saw Iraqi society "slowly coming back to life" with markets filled with shops and shoppers—something that would have meant risking your life only a few months earlier. The security forces we were training were finally performing, and, in a significant development, it appeared that the brutal tactics of al-Qaeda were finally being rejected by a majority of Iraqis. The terrorists had overplayed their hand, and people were sick of their bullying ways. The two scholars ended their op-ed by saying that "there is enough good happening on the battlefields of Iraq today that Congress should plan on sustaining the effort at least into 2008."

Colonel Peter Mansoor, who was serving as a top staff officer to General Petraeus at the time, remembered the op-ed's significance. In his book Surge: My Journey with General David Petraeus and the Remaking of the Iraq War, *Mansoor wrote that the essay "was nothing less than an intellectual bombshell." It changed the entire debate, he said: "Politicians who had been tripping over one another trying to figure out how to force the Bush administration to withdraw from Iraq were now content to await the testimony of General Petraeus and Ambassador Crocker."*

While our politicians and journalists argued in Washington, D.C., and while our commanders issued orders and made assessments in Baghdad, our warfighters executed the strategy in villages and cities throughout Iraq. One of them was a young paratrooper named Jeremiah Church. He grew up watching cowboy movies with his father on weekend afternoons, and their brand of frontier justice offered at the end of a six-shooter appealed to him from an early age. Church would play pretend army in the woods near his home in Gerry, New York, a small town about twenty-five miles east of the Lake Erie shoreline. He eventually moved from westerns to watching war documentaries on the Military Channel. "It's hard to explain why, but combat was something that I knew I would be good at, and I knew that at a very young age," he told me.

JEREMIAH CHURCH: I walked into my recruiter's office when I was nineteen years old and said, "Look, I want to be a paratrooper." The recruiter said, "I don't know if I can do that for you." I said, "All right. If you can't help me then I can't help you." I turned around to walk out and the recruiter said, "Whoa, whoa, whoa." He asked me to come back and sit down, and we started to talk.

I knew that I needed to get to Fort Bragg because I wanted to be in a rapid deployment unit so I could get into combat as soon as possible. There was a "wish list" where you could pick three places in the Army where you want to go. I wrote, "Fort Bragg, North Carolina. Fort Bragg, North Carolina. Fort Bragg, North Carolina." All the way down the whole list. For me, airborne was the only option.

There are all kinds of airborne personnel. You have your basic airborne infantry, and then you have mechanics, artillery guys, and everything else. I was a 19-Delta, a cavalry scout, and our numbers are less than that of an infantry platoon. We're supposed to have twenty individuals in our platoons, but you're lucky if you have that, compared to the forty-plus or so who are in an infantry platoon. That was one of the things that made it attractive to me, as well.

Tell me about your engagement in August of 2007, during the surge.

I was a specialist at the time, an E-4, and was about twenty-one years old. I was serving as a reconnaissance platoon machine gunner with Troop B, 5th Squadron, 73rd Calvary Regiment, 82nd Airborne Division. My unit was something called an RSTA Squadron, which stands for Recon, Surveillance, and Target Acquisition. They took platoons of infantry and platoons of scouts, and they stuck us together into one unit for the first time. The reason for that was we had a huge area of operations to cover in Iraq. We had to change the way we fought. We were the very first RSTA Squadron, to my knowledge, to deploy into combat. We did very well while we were there.

We'd been in Iraq for a little more than a year by that time in August of 2007. We had gotten extended and were on a downward slope, just about to leave. We'd spent the previous year on little patrol bases in the middle of nowhere, all on our own, out on the ragged edge, and they had started drawing us back. So by August they pulled us all back to Forward Operating Base Warhorse. There were some other people spread out here and there, but I laid my head down to sleep in Warhorse, just outside of Baqubah, Iraq, about an hour or so north of Baghdad.

Baqubah was one of the biggest cities that I had spent time in. It was very urban. The first time we went through the city it was like cruising through a ghost town. It was desolate, destroyed. My first impression was, "Holy shit," because the place was so ravaged by the war. There wasn't a wall that didn't have bullet holes in it. Windows were blown out and bomb craters were everywhere.

The army did a big push through Baghdad before we had gotten there, and they pushed south to north through Baghdad, up through the Diyala River Valley, and so all of the bad guys ended up flowing to the next biggest city for a safe haven ... and that was Baqubah. That's where some hard fighting had been. It was pretty well destroyed when we got there.

The unit before us had taken a hard licking. They pretty much sheltered up behind the walls of Warhorse and stayed there. When we showed up at our patrol bases we found that nobody was there. We were like, "What the hell's going on ... where is everybody?" Well, that's because they decided it would be safer to fall back to the big forward operating base. They pretty much abandoned a majority of the area of operation. The bad guys took over and held it. So, during the surge, it was our job to take everything back.

As a cavalry scout, airborne reconnaissance, our specific role changed depending on where we were and what we were doing. We spent most of our time flying around Iraq, getting in and out of helicopters. It was our job to seek out and find the enemy, and then pound them back into the Stone Age, wipe them off the face of the planet. We spent a lot of time sneaking around in the middle of the night looking for the enemy. We'd find them, call in additional forces, and then we'd rub them out. That was our job. Most of it was, "Take ground, take ground, take ground. Push, push, push," the whole time. That's what we did. We lived it.

I was a machine gunner almost the entire deployment. Before I would get on the helicopter, I would prepare by picking up my M240 Bravo and counting all my ammo. I usually carried about 1,000 to 1,800 rounds between my assistant gunner and me. We'd load all the ammo up in one great big belt and put it on our backs. I'd pick up my machine gun and often get on a helicopter at 2:30 in the morning. We'd fly to the city and get off in a heartbeat. As soon as you were boots on the ground, you were fighting. There was no waiting. We'd push all the way through the city until we had it secured, and then the helicopters would come pick us up. We'd get on them, and we'd go to the next town. We did that the whole year. We pushed those guys up out of the Diyala River Valley ... and then all the way to Iran.

Jeremiah Church behind the trigger,
Iraq 2006 (*J. Church*)

August 8, 2007, was an interesting day. It was supposed to be an easy humanitarian mission. The Iraqi police had given us reports that the insurgents had dammed up a canal system that provided fresh water to a farming village called Naqeeb, Iraq, which was outside of Baqubah. Our mission was to follow the Iraqi police officers to this village, find the dam in the canal, and then restore water flow to the village.

We travelled on Humvees rather than helicopters because it was so close to our base. I was going to be the mounted machine gun guy that day and I remember the night before thinking, "Man, I better make sure all these guns work." I don't know if it was premonition, but I had this gut feeling that those guns needed to work.

I knew my M240 Bravo was good to go because I'd been carrying that around everywhere I went for a year. It was my baby. I had been a 50-caliber machine gun guy before, but the particular 50-caliber machine gun that I was taking on the mission was new to me. I had not used it before. So I spent all night tuning, tweaking, filing, and getting it running good, which was a good thing because I sure as shit needed it the next day.

I took my M240 Bravo machine gun and mounted it on the outside of the turret on my right. Then I had the 50-caliber main gun mounted center on the turret. We left the base in a convoy of eleven gun trucks and rolled into the village probably around ten o'clock in the morning. The village was full of your standard Iraqi mud huts, and the area around the village was agricultural. There was also a giant canal system feeding the farmland. Driving along the canal roads is what made it so bad. There was water all around, so we couldn't maneuver. We couldn't back up, couldn't turn left or right. Once you were up there on the canal road, you were stuck. It was hot, probably 140 degrees. It felt like somebody was blowing a hairdryer in your face.

We got to the spot in the canal system where the dam was supposed to be ... but it wasn't there. There was no dam. Nothing. So we asked the Iraqi police, "Where's this dam?" They said, "Oh, it's right up here." We moved up, but again, we saw nothing.

In my opinion, the story about the dam was total and absolute bullshit. I can't confirm it, but I honestly believe they were baiting us into an ambush. I just had a bad feeling, so I said to my commander, "Sir, shit ain't right here. This isn't a good idea."

I don't know what our leadership was thinking, or what other information our company commander had, but from my position— and having spent most of the deployment in the field fighting these guys—it all seemed like an obvious setup.

Then the Iraqi police said, "Hey, we know where this insurgent has a DShK heavy machine gun. Do you guys want to go get it?"

Now, that's a serious gun, a Russian-made weapon that can do some real damage. It's quite a little bit bigger than our American 50-cal. It shoots a bigger round. Nasty, real nasty. Armor-piercing, incendiary rounds. Not something you want to mess with. So everyone was like, "Well, shit. We're here. There's no canal blockage. They say a DShK is just up the road. Screw it. We'll go get it."

I guess that was the consensus. But what we didn't know was that the insurgents had that DShK ready, mounted on the back of a pickup truck, and were getting ready to put the boots to us with it.

Anyway, we followed, and the Iraqi police kept egging us on, saying "Hey, let's go a little farther." Every time we would stop, I'd think, "Something ain't right." That's when the Iraqi police would say, "No, no, no. Let's go a little farther." So we followed them a little farther, and eventually we followed them right into an ambush.

I was the gunner on the lead truck and could see all the signs. There were no people around, and no kids, and that's a big thing. There should have been kids asking for pencils, and candy, and bottles of water. But they weren't there. That was a big red flag. Then I started noticing that some of the buildings to my right had bricks pushed out of them, what we call "murder holes." Firing ports, so they can shoot out of the walls.

Me, I worry. We should have stopped right there. We should have known better. But we were told to push for whatever reason, so we pushed.

Then I noticed, no shit, full-on sandbagged machine gun positions on top of the buildings: machine gun nests with intersecting sectors of fire. I radioed back what I saw, and was told to take a picture of the fighting positions and push ahead. So ... we pushed ahead down the canal road. If you were looking out of the front of my gun truck, all along the right-hand side of the road was a water-filled canal and then small buildings. To my left was a big drop-off, then a flooded field. In front of us, down at the end of the canal road, was a T-shaped intersection.

We were moving on this high canal road when I noticed, down in the tall grass right on the side of the road, an asshole with an AK. I called it up and the commander said, "Smoke them." So we initiated fire on this guy, and the minute we opened up it seemed like the whole fricking world exploded. There were bullets flying out of every window. From every ditch, every nook, cranny, and rooftop. There were so many muzzle flashes that you couldn't even count them. Imagine dropping a handful of rocks on a piece of tin, then letting them slide down the side. That's what my truck sounded like. It was getting hammered so hard.

There weren't many spots where they couldn't hit us. They also had trenches cut between the buildings so they could go from building to building without being shot. These dudes were pros, not your basic goat herder with an AK. These were trained professional fighters.

We had a canal on one side, a flooded field on the other, and behind us was our entire platoon stacked up on the road. I started engaging to the right with my 50-caliber machine gun. Those guys were close. Real close. During my previous firefights there were times when I was shooting at only figures in the distance, outlines of bodies and AKs. These dudes, though, I could see their fricking teeth and the whites in their eyes. I kept engaging guys on the rooftops, in the windows, and dudes running between the buildings where the trenches were cut. I could only see them from the shoulder up, so I tried to zip those guys as they were going back and forth.

Then I hear this strange "whoop, whoop, whoop" deep machine gun sound on a very slow rate of fire. Then I start seeing big green golf ball-sized tracers rip past my truck. I turned back to my left, to the twelve o'clock position of the vehicle, and then looked down the canal road. There, blocking the only direction we could escape through, was a pickup truck pulled across the street. In its back was the DShK, mounted with an insurgent firing at us. In the back of the truck there were four guys: the machine gunner with the DShK, an assistant gunner, a guy feeding ammo to the machine gun, and a guy with an RPG tube. I started engaging them immediately, probably piling one hundred rounds or so into that truck. I gave them the pink mist treatment.

The what treatment?

We called it "pink mist." When you hit a human being with a 50-caliber machine gun, they tend to split in half and partially vaporize. It makes a big red mess; a pink mist. After I took out the crew, the driver hammered the gas and drove the truck into an alley, out of view.

After taking out the truck, I swung back around to my right-hand side and started engaging targets in the buildings again, the guys that were real close. I remember trying to pick my targets. This wasn't a "spray and you're going to be fine" situation. This was a "dude, you need to make this ammo last" situation.

Even though we had thousands of rounds on that truck, the volume of fire was so heavy that I knew it wasn't enough. We could not have had enough ammo on that truck to sustain a firefight that big. It was either pick your shots and last as long as you can, or waste all your ammo and get overrun. I was trying to be as meticulous as I could about where I was shooting and who I was shooting. No room for waste.

We were taking fire from all sides. It was pretty nasty. I was getting hit from the front, from the left, and from the right. I would pick a target, then transition—from the twelve o' clock, over to my three o' clock, back over to the nine o' clock, and so on. I was picking the dumb guys that were repeatedly popping out of the same windows or same parts of the canal. They weren't moving around a whole lot, so they were easy pickings. The majority of the entrenched positions were to my right. That's where they had the machine gun positions, the trenches cut, and the fighting positions on the rooftop, so that was my priority because they could fire right down into my turret.

At that point I realized ammunition was getting critical. I was probably a couple thousand rounds into the firefight by then, between the 50-caliber machine gun and the M240 Bravo machine gun. Then I ran completely out of 7.62-millimeter for the 240, so I was sticking mainly to the 50-cal.

Meanwhile, the driver of that truck with the DShK had gotten a whole new gun crew and they were back in the same intersection, blocking our path and putting fire on us pretty hard. Like I said before, the DShK is a very big, very nasty machine gun. The tracers flying at us looked like green glowing ping-pong balls. They're huge. At that point I remember thinking, "I'm done playing this game." I probably pumped another 250 rounds of 50-caliber into that truck. I just pumped and pumped and pumped the rounds until my gun went "click" and there was no ammo left.

I remember looking over the barrel and down the street; I had pumped so many rounds and tracers into the DShK truck that it was on fire. I remember thinking, "OK, that's taken care of," and that's when I felt something slap the shit out of my left wrist. Like someone just slapped it as hard as they possibly could. It tore my hand off of the gun. I'd never been shot before, and I didn't know what happened. I remember thinking, "What the hell was that?" I rolled my wrist over and realized I had a pretty substantial injury. It was like a sprinkler. Blood flew everywhere.

I remember yelling down into the truck, "Holy shit! Hey, I'm hit. Give me a damn tourniquet!" One of my buddies, Tyler, who had been in the back seat passing up ammo and engaging out of the window helped me get the thing tourniqueted. Then I climbed back into the turret and started shooting again.

Your left wrist was wounded, so were you shooting with only one arm?

The beautiful thing about that 50-cal is that it's ambidextrous. It's got a butterfly on it, not a trigger. You can hold it and fire it one handed. Obviously you're not going to be as accurate, because that gun jumps around a little bit. But I had other problems. Have you ever been ridiculously dehydrated, where your body is just kind of super weak? That's how I felt. I didn't realize that when we put that tourniquet on that it was too close to the wound, so the artery had retracted and I was still bleeding profusely. In fact, I was bleeding to death and didn't really know it. I kind of slumped down inside the turret and slid into the back seat.

My buddy Tyler was like, "Dude, are you good?" I said, "Yeah, I'm good." Tyler smacked me on the helmet and then he climbed in the turret and started firing.

I put on another tourniquet, then another, but I was still bleeding. Then I put on an Israeli pressure dressing on top of both tourniquets, and the blood flow slowed enough to where I wasn't worried about it that much. It was kind of oozing, dripping and seeping. But at least it wasn't spraying. I think I passed out for a moment, as well. After that, I got myself composed and thought, "OK, I'm good to go. I don't have anything to worry about." Then I heard Tyler's gun stop firing. I look up in the turret to see what was the matter. Tyler looked down and said, "I need 50! I need 50!"

I looked around in the truck but there wasn't anything but empty ammo cans.

I yelled, "We don't have anything left!"

"Well, give me 7.62!"

"We don't have 7.62, man. We're black on ammo!"

Then I got an "Oh, shit," sense of dread because the position of our truck on the road meant that we were the only people who were able to effectively return fire on the enemy. If we were out of the fight … if we were done for, then we all were done for. There was nobody to hold back the fricking horde from coming over the wall.

You know how, back in the day, everybody had rifle racks in the back window of their pickup trucks? Well, we kind of had the same thing. We rigged up an M249 light machine gun to hang on the steel bulkhead that separated the cabin from the trunk. It was just hanging there, like a redneck rifle rack. The 249 is a tiny thing and fires the same rounds as an M4, like an AR-15, but it's belt-fed so it has a decent rate of fire, but not the stopping power of the larger guns in the turret. But still, when I remembered it was there, I was like, "Thank God, we've got something." I cut the paracord holding the 249 against the bulkhead and threw it up into the turret. Tyler started engaging.

Now, that weapon was there for absolute emergencies. We'd only kept 1,800 to 2,000 rounds for it. That's a minute and a half, two minutes, maybe, in a firefight that heavy. That ammo wasn't going to last. So I threw as much belted ammo up in that turret as I could. Then I grabbed my M4, a couple of magazines, threw a mag in it, racked a round and dove out of the vehicle. I ran to the back of the gun truck, ripped open the trunk and started grabbing boxes of ammo while still shooting.

How, with a wounded wrist?

I laid the M4 across the crook of my left elbow. That's what I was using to stabilize my rifle so I could shoot accurately, rather than one-hand Rambo blasting. I was still taking accurate shots. All we had in the back of the truck was belted 5.56 ammunition, maybe a couple boxes of 7.62. I grabbed it all and handed it up to the turret. Then I stood by the corner of the truck, engaging targets that were right there in the canal. These guys were close, fifteen feet to twenty feet, maybe. I could see their facial expressions, the look in their eyes. I think when Tyler stopped shooting they figured we were done, and when I was out of the truck they figured I was easy pickings. They were probably going to try and grab me, take me off as a war trophy.

They were trying to get across the canal, but I was still shooting. Tyler reloaded and laid down some good bursts of fire, too. Then I ran to the back seat of the truck and flipped the door open, using it for cover. Then Tyler ran out of ammo again. I thought, "Screw it." I grabbed some mags, reloaded my M4, and ran to the gun truck behind us to get their ammo. They supplied us with some 50-cal and some 7.62. I ran all that back up to my gun truck and dove into the back seat, *Dukes of Hazzard* style, and started handing ammo up to Tyler and kept shooting out of the open door. I remember standing at the back door of the truck firing rounds and screaming for magazines. Somebody would hand me a magazine and I would fire, fire, fire. Tyler was blazing away up in the turret.

My lieutenant was a replacement for the one we had deployed with, so this was the first firefight we had been in with him. We hadn't been on many missions together, but the dude knew what he was doing. He had his shit together. He screamed, "Church! Get in the truck! We're getting the hell out of here!"

So I sat up in the truck. The driver, David Holtz, slammed the truck in reverse because the insurgent's vehicle was still in front of us in the intersection on fire, so it was impassable. As we backed out of there we were still getting slammed. I remember Tyler yelling, "They're coming! They're freaking coming!" I looked out of the window and those dudes were chasing us down the street on foot, trying to catch up as we were backing up. Tyler was mowing them down and I was pumping lead into them as we reversed.

We got backed up into the main section of town, where a big market area was. We circled all the gun trucks up, our platoon and then another platoon that came up behind us. I guess the bad guys didn't want to play with that many gun trucks together, all brimming with 50-calibers and Mark 19 grenade launchers. I don't think they wanted to push their luck, because they didn't have the advantage anymore. They just stopped. I don't know what the hell they did after that, but they stopped shooting at us and we stopped shooting at them. They literally disappeared. It was creepy.

I had gotten out of the truck and I was like, "Son of a bitch!" I ripped my helmet off and threw it up against the side of the truck. I said, "What the hell are we doing?" I was pissed because we had backed out of the fight. After I had gotten shot, I don't remember being like, "Oh, no, I'm going to die," or anything like that. Instead I went in 100 percent, full-on rage mode. So I was really mad that we had backed out of the fight. I didn't realize that nobody had any ammo. I was the only casualty, but the trucks were barely running.

They were so shot-up we had to tow a few of them back. They were all Swiss cheese, mutilated, wrecked. We couldn't have kept fighting, but I didn't realize it at the time.

I had slumped down to the ground by the right rear tire of my gun truck. Guys were running around trying to get trucks running and trying to scrape up any ammo they could find. The medics started working on me. I had a new doc that had never dealt with a combat-related injury up until then. He was messing it up. I cursed him out because instead of giving me morphine the correct way, he ended up pushing it into the IV bag instead of into the injection port. It squirted all over the side of the truck. I can't blame him, though. He was a replacement and had only been with us for a little bit. He came straight out of the States into Iraq, into our gun truck, and then into an ambush. We had just gotten the snot pounded out of us and he was as hyped and adrenaline-rushed as I was. His hands were shaking. Luckily, the second platoon had their doc and he knew what he was doing. He took over, calmed the new medic down, and showed him how to handle the wound.

While they were working I asked my guys, "What the hell are we doing? We've got to get back in there!" One sergeant said, "No, you're not going anywhere. You're going to die if you don't get treated."

There were roughly fifty of us, and the intelligence reports that came back after the ambush said there were about 150 enemy fighters. We were outnumbered three to one, at least, if not more. We also didn't have any aerial support or any indirect fire support because we were a few kilometers outside of our area of operation. We didn't have the right radio frequencies for that area.

We finally got ahold of a medevac but they wouldn't bring in the helicopters because it was still a hot landing zone. We were all trying to figure out what to do and then I finally said, "Why don't we just get in the trucks and get the hell out of here?" Everybody looked at me like, "Oh, no shit." So we piled into the trucks and attached tow cables to the ones that weren't running and hauled ass back to the FOB.

Was there a hospital on the base?

Yes. We were on a big base, and it took a while to get from the gate to the Level 1 care facility. They put me on a stretcher and wheeled me into the room. Doctors were running around everywhere, screaming for meds, and this and that. I was lying on my back and they had my left arm raised above my heart. This doctor kept messing around inside my wrist with a pair of pliers or whatever. It hurt, so I was screaming some pretty crude stuff at him.

Every time he would go in there to try to pinch the artery off, I would scream and yell. I eventually ripped the railing off of the gurney. Then some huge male nurse came over and said, "Hey, if it hurts you can squeeze my hand." He put his hand in mine and the minute the doctor went back into my wrist it hurt like hell, so I squeezed the nurse's hand.

Now, I'm not a burly guy at all, and this dude was probably twice my size. I think it had something to do with all the adrenaline, but when I squeezed his hand it felt like I crushed a Dixie cup. Just crunched it. I remember him screaming, "Ouch! Let go! Let go!" So I let go. I remember somebody looking at him and saying, "Bad idea." He walked away to get his hand looked at. I found out later that I had broken a couple of his fingers.

My squadron commander and my sergeant major came in and tried to calm me down a little. The nurses were cutting off my ACUs [*Army Combat Uniform*]. I didn't really care about the uniform, but when they got down to my boots, which were brand new and I really liked them, I screamed, "don't cut off my boot laces! I'll kill you if you cut my boot laces. Untie them instead." The nurse looked at me, looked at the doctor, then the doctor looked over at my commander and asked, "Are all your guys like this?" By that time I had gotten a reputation for being a little bit wild and aggressive. The guys in my platoon even nicknamed me "Manimal." So my squadron commander looked at the doctor and points down at me and said, "No, just this one."

So they were working me, pumping fluids and blood into me, and I remember bragging to my commander and sergeant major about taking out that DShK ... then I realized they were probably going back out without me. I was still lying down, and I grabbed my commander's coat, "Sir, I know you guys are going back out there, but I'll be back in a couple of weeks. If you guys go without me, I'm going to be seriously pissed."

In my head, I was still thinking that I was fine and would be right back in action. My commander looked down, glanced at my injury, and said, "No. I'm sorry, son, but you're going home." [*Church pauses*] "You're going home." That's not what I wanted to hear.

Church was credited with destroying the DShK machine gun, covering his platoon during a significant firefight, and killing at least eleven insurgent fighters. For his actions he was presented with the Silver Star.

Meanwhile, back home the partisan political debate intensified in anticipation of the assessment that would be delivered in early September by General Petraeus. Republicans who supported the surge were feeling confident due to the relatively positive reports coming from Iraq, but it had been such a long, hard, and costly war that neither victory in Iraq nor support from the Congress could be counted upon.

A few days after Church's battle, neoconservative writer Bill Kristol spoke for many surge supporters when he penned an article titled "The Turn: Defeatists in Retreat," in the Weekly Standard: "The Republican party faces a moment when, to paraphrase Winston Churchill, honor points the path of duty, and the right judgment of the facts reinforces the dictates of honor. General Petraeus will deliver the facts in September. If Republicans can keep their nerve under media and elite assault, then they will have the honor of following the path of both duty and the right judgment of the facts. I suspect all will come out well. Americans can sometimes be impatient and shortsighted. But when a choice is clearly presented, they tend to reject the path of defeat and dishonor."

Indeed, Americans were impatient, but many were reading about the surge and daring to hope for the best.

Jeremiah Church is presented with the Silver Star,
November 2007 (*U.S. Army*)

CHAPTER TWELVE

"THEY CALLED THE NEIGHBORHOOD 'AL-QAEDA'S CASTLE.'"

ERIC GERESSY,
SILVER STAR, *BAGHDAD*

As additional U.S. troops flowed into Iraq and their new strategy successfully pushed the insurgents farther away from their safe havens, the terror networks became increasingly brazen and brutal. They knew their grip was weakening, and that they were in danger of being broken, scattered, and defeated by the coalition's new counterinsurgency campaign and the Anbar Awakening movement. In a nation where suicide bombings were becoming part of the daily routine, a coordinated attack in northern Iraq in mid-August 2007 would remind everyone that civilians still weren't safe, and the war was far from over.

"Five fuel tankers were driven by suicide bombers into two crowded villages belonging to Kurdish members of the Yazidi religious sect before they were detonated," read an article in the August 16th edition of the Telegraph. "A vast number of clay-built homes ... were leveled by the blast on Tuesday night which was followed by an enormous fireball." At least 250 were killed in the cowardly attack and more than 350 more were wounded. Nearly all were civilians—women, children, babies, and the elderly.

The insurgents' desperation showed in their attempts to foment sectarian violence and undercut the surge's progress. But despite al-Qaeda in Iraq's best efforts and worst atrocities, our warfighters had seized the initiative. In late August, the radical Shiite cleric Muqtada al-Sadr suspended his Mahdi Army after it clashed with government forces, a *welcome sign that both Sunni and Shia militias were finally standing down.*

By early September 2007, the number of American troops in Iraq reached its high water mark at approximately 168,000. Many believed it was still far too few, but it was certainly more than the 127,000 who had slogged their way through a disastrous 2006. But as proponents of the new strategy said, our counterinsurgency campaign wasn't just a surge of troops; it was a surge of ideas. Our men and women in uniform had always done things right, but now they were finally doing the right things.

Every defense policy maker in Washington, and certainly our enemies in Iraq, eagerly awaited the testimony that Petraeus would deliver to Congress beginning on September 10 in a joint hearing of the armed services and foreign affairs committees of the House of Representatives, followed by a similar gathering the next day in the Senate. It was make-or-break testimony, and the president and Congress would take action based on whatever the general said, and even how he said it. The context surrounding the hearing was crucial. Supporters of the strategy pointed to evidence of success, like the trends highlighted in the op-ed written weeks earlier by the two Brookings Institution scholars. Opponents pointed to the headlines still showing U.S. troops being killed and innocent Iraqis being slaughtered.

Even though our politicians remained unsure, evidence in Iraq indicated that al-Qaeda's leaders knew they were losing. They didn't need an analytical assessment from political leaders to feel the impact of the daily losses. They were on the run, everywhere, and our warfighters were denying them any safe haven. They were losing on the real battlefield, so the insurgent leaders turned to the only battlefield in which they continued to enjoy some success: the media.

Al-Qaeda figured that if it delivered a spectacular blow against coalition forces only days before Petraeus delivered his testimony, any positive trends he could highlight as justification to continue the surge would be lost amid the news surrounding a catastrophic loss of American life. If they could wipe out a platoon, for instance, or maybe even an entire company of soldiers or Marines, then nobody would listen to the metrics and statistics showing that the surge was having an impact. Our already wobbly politicians would say the testimony given by Petraeus was questionable and calls for a complete withdrawal would gain serious traction before the general even left Capitol Hill.

So al-Qaeda found a target—a small base on the outskirts of Baghdad called Combat Outpost Blackfoot. A new company of soldiers would be assuming command there in early September, just days before the Petraeus testimony was scheduled. It'd be a bunch of new guys, green and untested. Al-Qaeda thought if they massed a company-sized element of fighters, about two hundred or so, and attacked the Americans before they had a chance to establish their defenses, then they could deliver the much-needed attack that would derail Petraeus's testimony and send the American public clamoring for a withdrawal. It was a sound plan by our enemy, yet the insurgent leaders didn't know that the new company of American soldiers had a combat-tested first sergeant named Eric Geressy.

This was Geressy's third deployment to Iraq. He told me that his second tour, from 2005–2006, was a "nightmare" and that he felt "done, both physically and mentally" after it was finished. The memories of the carnage surrounding Baghdad's hospital complex, and pulling small children from the burning vehicle, were only a few of the many things he tried to forget. He and his wife divorced within a month of his return home, and the career soldier, then thirty-six years old, wasn't even sure he wanted to be in the Army anymore. The only thing Geressy did know was that he needed to go someplace different, and do something different. He called the office that handled assignments and learned that there was a first sergeant opening in a Stryker unit in Germany. He hadn't spent any time in Germany since most of his time had been with airborne and light infantry units, but it was the change of pace he needed. So, within thirty days of coming home from combat, Eric Geressy found himself headed to Europe. He didn't expect to see Iraq anytime soon, and was thankful.

ERIC GERESSY: I arrived in Vilseck, Germany, and reported in to Command Sergeant Major Victor Martinez, the regimental sergeant major for 2nd Stryker Cavalry. We talked for a while. He was a great command sergeant major who really cared for the troopers of the regiment. He then had me assigned to the Headquarters Company for 2nd Squadron, 2nd Stryker Cavalry Regiment—the Dragoons. Then, after about a week in Germany with my new unit, we received orders to deploy to Iraq as part of the surge.

It seemed like things had gone a hundred miles an hour. I was in our unit motor pool while the squadron command sergeant major, Frank Wood, had all the troopers reciting the Dragoon Creed at the top of their lungs, getting them motivated for the deployment. That's when it all hit me at once ... everything that had happened in the previous fifteen months—the hard deployment, the divorce, the move. I just wasn't sure if I could do the job anymore. I wasn't sure if I was ready to go back to Iraq.

Immediately after the formation I went in and told the headquarters company commander that I didn't think I was ready for all of that again. We then went to see the squadron commander, Lieutenant Colonel Myron Reineke, and also Command Sergeant Major Wood. I explained the situation, and how I felt at that point. Lieutenant Colonel Reineke told me to take it easy for a few days and think about what I wanted to do, then come back to see him and the command sergeant major and we'd talk.

I went home that night and sent a few emails. One email was to Colonel Mike Steele, who was my former commander in Iraq during my 2005–2006 deployment with the 3rd Brigade of the 101st Airborne Division. Another email was to my old battalion executive officer, Major Steven Delvaux. I explained what was going on, and they both contacted me immediately, and provided very similar advice.

It was simple, they said: if it was time for me to retire and hang it up, that was okay because I had already done my part. But if I felt I could still contribute to a unit going into combat, then that's what I should do. Colonel Steele then reminded me that if he could, he would have wanted his son to serve with me. Those were strong words coming from him ... and they made my decision very easy.

I would be going back to combat.

I learned a lot after two combat deployments. I knew the fight, and I felt I could make more of a difference if I was placed back in a rifle company. I talked to some folks about it but they told me that

I'd never make sergeant major without spending time in a headquarters company. But, for me, it wasn't about making rank. It was about going where I felt I should go and where I could do the most good. They counseled me against it, but I had made up my mind. You see, a rifle company's job is simple: close in on and destroy the enemy. That's it. When most units go out, they're looking to go from point A to point B without any issues or incidents. Infantry rifle companies are the only guys going out there looking for trouble. They're looking to make contact with the enemy, not avoid it. They're looking for a fight. Everybody in the Army has an important role, but that's what separates the infantry from everybody else.

The next day, I went to see Lieutenant Colonel Reineke and Command Sergeant Major Wood, and I told the commander, "Sir, I'm ready to deploy again, but I don't think being a headquarters company first sergeant is a good fit for me. I want to be back in a rifle company." I got what I asked for, and was re-assigned to be the first sergeant of Eagle Company, 2nd Squadron, 2nd Stryker Cavalry Regiment—the "Second Dragoons."

Were you even allowed to return to combat so soon? I thought there were rules against that.

The Army had implemented a requirement for soldiers to spend at least twelve months back home before going on another combat deployment. It was called "dwell time." I didn't actually have to go with the Dragoons to Iraq since I had only been back from combat for about two months when my new unit received the deployment orders. So I had to sign a waiver stating that I wanted to go back voluntarily. Even though I wanted to, I remember signing the waiver request and then briefly thinking, "What the hell am I doing?"

What that a hard decision?

It sounds difficult, but after seeing those young and inexperienced guys getting ready to head into Iraq, deciding to be with them was a pretty easy decision. Somebody needed to take care of them, right? After two years of combat in Iraq, I felt I knew the nature of the fight and that I could pass that experience on to these troops. If I was able to help one soldier get back home safely to his family, then it would all be worth it to me no matter what. After I signed my waiver, several soldiers who were also inside of their dwell time from previous combat deployments walked into my office and said, "First sergeant, we're going to sign our dwell time waivers, too, because you did it. We want to go with you." At that point, I felt 100 percent sure that I did the right thing. I thought, "They're going. I'm going. Yeah. Let's do this."

My first two deployments were still fresh on my mind and I had a clear vision for how we needed to train the soldiers for what we were going to face in Iraq. I pushed them hard while we were training in Germany. Real hard. I had them doing long runs in the morning and ruck marches with all of our gear in the afternoons—full pack, full kit. We trained on our weapons, our small arms, and our machine guns. The standard for every Eagle Company soldier was to qualify "expert" on their assigned weapon. We ran battle drills. All day long you'd see my company out there. People knew we were preparing.

About two weeks before we deployed from Germany, we got orders for our location. Our squadron was going to the outskirts of Baghdad, to Forward Operating Base Falcon. I was there in 2005 with the 101st Airborne, and I knew it was going to be tough ... I never forgot all the memorial services I attended when I was there. I kept that to myself and just answered, "It was tough," when asked. I knew the company would be ready.

When we finally got to Kuwait we had a few weeks for some final training at Camp Buehring. We turned up the training tempo. I took advantage of the weeks we were in Kuwait to get them ready for fighting in the extreme temperatures we would face once we got into Iraq. It was still summer, and the change in weather couldn't have been more different from Germany to Kuwait, where it was maybe 110–120 degrees Fahrenheit in the day and ninety degrees at night. Most of the other companies were staying inside because it was so damn hot. Not Eagle Company. Not us. I was very no-nonsense, and I wasn't messing around. By the time we got to our outpost in Baghdad, everybody had about enough of me. I knew that, but I wasn't competing in a popularity contest. I wasn't going to lose a soldier because they weren't prepared for the enemy, the heat, or the rigors of combat.

By the time we were finished in Kuwait the soldiers of Eagle Company, 2nd Squadron, 2nd Stryker Cavalry Regiment, were ready for what they would face in the next few weeks. The soldiers all knew that they had the best guys training them, the best guys leading them, and when it got tough, they could count on those on their left and those on their right. Our squadron commander and command sergeant major knew they could count on Eagle Company, too, so we were given the mission to occupy a combat outpost in East Rashid, Baghdad. This would be the most dangerous location the unit would be in during that time, and it was the squadron's main effort while in Baghdad.

Was Baghdad the same as you remembered?

No. It had undergone a tremendous change. The surge had been going on throughout the summer and the enemy was boxed into little al-Qaeda strongholds. They were dug in and holding on. They were way more organized as well. In '05–'06, al-Qaeda was fighting in squad-sized units, six to nine men at the most. They'd initiate contact and then quickly run away. Every contact with the enemy was very violent, but normally did not last very long. In '07–'08, though, al-Qaeda was maneuvering in platoon-sized elements and even company-plus sized elements on occasion, which is about two hundred guys. Their attacks weren't over in five minutes, like before. During that time period in our sector, the al-Qaeda in Iraq fighters would maneuver on your position, then stay and fight. Sometimes they would go on for an hour or two. It was crazy. I couldn't believe it. They wouldn't break contact. They would maneuver. They would mass. They would isolate us. I was shocked at the conditions at that time. I was surprised how organized and how determined al-Qaeda was to take the fight to us. It was a totally different situation than I experienced in my previous combat deployments.

We conducted what's called a "leader's recon" to our new post; this included all the company's platoon leaders and platoon sergeants, along with members of the unit we would be replacing. We got to see the neighborhood we'd patrol and the base that we would call home—Combat Outpost Blackfoot. The unit we were replacing had taken a lot of casualties. They had a lot of guys wounded. They had several guys killed. It was a tough fight for them, and by the time we would occupy and relieve the unit at the outpost, they were only manning the position with one platoon, which was about thirty-five soldiers. They didn't patrol in the daytime, and from what I could tell, they focused mainly on securing the outpost.

The outpost itself was not built up into the fighting position that I would have expected, and I took note of that. Also during the leader's recon, while on the roof of the outpost, I remember seeing a big piece of steel, maybe two hundred meters from the entry control point north of the outpost. I asked their commander, "What's that?" He said, "That is part of one of our Strykers. It got blown up with an IED." I could not believe that they got hit with an IED only two hundred meters from the outpost, right out in the open, and nobody saw the IED being put in place. I thought to myself, "That isn't going to happen to us."

After the recon was complete, we went back to Falcon and briefed our squadron commander and the rest of the company. Lieutenant Colonel Reineke, he asked me, "So, what do you think?" I said, "Well, sir, I don't think we are going to have a problem finding the enemy." He asked, "What do you mean by that?" I said, "Well, apparently they are right at the front gate to the outpost. There's half a Stryker blown up out there, and the unit we were replacing just left it blown up in the street."

I started thinking about everything that needed to be done at that outpost. I just had this feeling that I needed to get out there immediately, establish the priorities, and get to work building a defense. I also asked our squadron commander for extra medical support. He allowed us to take a physician assistant, a PA, named Captain Benjamin Blanks, and a couple of extra medics. That proved to be a wise decision. That medical team went on to save many lives, both American soldiers and Iraqi civilians.

The three rifle platoons moved to the outpost first, while I stayed behind at Forward Operating Base Falcon to ensure our Stryker vehicles were ready to patrol in Baghdad. They still had a few things that needed to be done. With most of the company at the outpost, there were only a handful of junior enlisted soldiers to do the work on the Strykers. I was there to make sure they got it done, but they ended up not needing me; those young troops knew exactly what they had to do, and then they made it happen. Besides me, a sergeant, a corporal, and a bunch of privates through specialists began working on those vehicles. The squadron gave us priority for anything we needed, and in only two days we got three platoons' worth of vehicles ready.

Our new outpost, Blackfoot, was in a neighborhood called East Rashid in southwest Baghdad. During that time period, al-Qaeda had gotten beaten all over the place until they finally went into this neighborhood. They called it "al-Qaeda's Castle," their last stronghold in Baghdad.

The area was a mix of Sunni and Shia Muslims. The Sunnis were predominantly supporting al-Qaeda in Iraq, while the Shia were supporting a group called the JAM, the Jaysh al-Mahdi, which is the Mahdi Militia that fought for Muqtada al-Sadr. There was a lot of sectarian violence happening at that time, Sunni-on-Shia attacks. Iraq's prime minister, Nouri al-Maliki, had already started a pro-Shia campaign to go after the Sunni Iraqis. So the Shia in this neighborhood, East Rashid, were trying to wipe out the Sunnis, and the government forces were helping them. That gave al-Qaeda an opening. They came in and said to the Sunni residents, "Hey, we'll protect you guys and help you fight the Shia government and the U.S. Forces."

So the people allowed al-Qaeda to move in and take control of the whole neighborhood, but then they just ended up destroying what was left of the place. They were killing locals in what we called EJKs—extra-judicial killings. If the locals did something al-Qaeda didn't like, they were executed on the spot. If anyone broke the rules, they would get beaten, often beaten to death. Al-Qaeda was abusing a lot of their women, too. So eventually, al-Qaeda wore out their welcome. I think that it just all came to a head between the locals and al-Qaeda, who were mainly foreign fighters or from other parts of Iraq. That change started happening about the time we moved into the area. So we had a small window of opportunity to take the fight to the enemy, and then show the people that we were there to help.

Two days after the leader's recon, on September 3, we moved from Forward Operating Base Falcon to our new home at Combat Outpost Blackfoot. Most of my guys had never driven through Baghdad before, so we moved under the cover of night. We had three platoons' worth of vehicles—twelve Strykers. It wasn't that far of a drive but we went through some really bad areas. I kept thinking … if we got into contact with the enemy during that night, it would be interesting since that was the same group of junior troops who got the vehicles ready. I had confidence in them, though, so away we went. The movement went smoothly, minus a slight break in contact, but eventually we made it to Combat Outpost Blackfoot without any incidents—no contact with enemy forces, no IEDs.

Combat Outpost Blackfoot was an interesting place. It was housed in the complex of an old Chaldean Catholic Church semi-nary—the Pontifical Babel College of Baghdad. It had been around for centuries. We had gotten permission from the bishop of Baghdad to secure the place.

The residents of East Rashid were suspicious and afraid. When we first arrived, only a few Iraqi civilians would meet with us. They were Sunnis, but probably more former Baathist than al-Qaeda sympathizers. They were on the fence because they were fed up with al-Qaeda and they were reaching out to us, feeling us out to see if they had a chance of working with us to make things better. The other civilians wouldn't come anywhere near us during the first few weeks. On top of that, we couldn't go outside of the outpost without getting into a firefight. It was so bad that first couple of months that we were getting into three and four firefights a day. Some of the fights were quick, but most would go on for hours at a time.

The outpost complex had three different buildings that were not linked together. The first was the school. It was a large building with a chapel inside and a rooftop area. That was our main building. The other was a large church on the northwest corner surrounded by a wall. The third area was in the northeast corner with another church. It looked just like a Catholic church inside. We were told the priest had been murdered there on the Easter Sunday before our arrival ... it was still decorated for Easter. They just left it that way as the people fled during the sectarian violence that plagued the sector before our arrival.

Christian churches were a favorite target of al-Qaeda. "Chaldeans have been moving ancient artifacts and century-old manuscripts around the country in order to protect and preserve the items," according to one Chaldean priest from Iraq. "Priceless relics of first-century Christianity, books in Aramaic, journals, diaries, paintings, sculptures, and other pieces were lost when churches were firebombed and ransacked by Islamic terrorists." The priest added that, fortunately, the U.S. Army sealed and secured the library at the Pontifical Babel College of Baghdad, which contained many ancient writings, before they took the abandoned structures as a base of operations.

Overall, the neighborhood was a strange place. On one side were all the Shia Muslims. On the other side were all the Sunni Muslims, and then there we were, in the middle, occupying an old Catholic seminary school. They were fighting each other, and then they were both fighting us. It was a very bad situation.

I thought that all of the priests had left, but after being there for three months, and through several firefights, we learned differently. One day I took a patrol across the road and over to the northwest corner, near that other church that was part of the complex, and found an old lady and a man living there. He looked like an Iraqi Santa Claus, with a long white beard and everything. I was totally surprised. I asked him, "How long have you been here?" He said, "Oh, about ten years." I asked,

"You've been living here this whole time, through the fighting, too?" He said "yes." It was amazing. He had been getting food from the Shia and Sunni residents of the neighborhood. They all knew him and liked him, and even though all that sectarian fighting had gotten really ugly, they were still taking care of that old priest.

Combat Outpost Blackfoot, formerly a Catholic seminary, Baghdad, September 2007 (*U.S. Army*)

Eric Geressy stands beside statues of Jesus and Mary at Combat Outpost Blackfoot, 2007. (*E. Geressy*)

Anyway, getting back to our time in early September … as soon as our company made the initial trip from Falcon to Blackfoot, the company's leadership—the commander, executive officer, and the fire support officer—had to go to a meeting someplace in Baghdad. That left me as the only company-level senior leader on the outpost. I had a handful of guys from the headquarters section, and the three rifle platoons, which were each led by a lieutenant and a platoon sergeant. That was it, about eighty to ninety soldiers, and about three-quarters of the enlisted men, and all of the officers, had yet to serve in combat.

Immediately after arriving at the outpost, I had a meeting with the lieutenants and the platoon sergeants and established the priorities of work in order to build the defensive positions of the outpost. Some of them were fighting over who was going to sleep where, and what platoons were sleeping in what rooms. I explained to them that what rooms they would sleep in was not the priority. I said, "The priority right now is to turn the outpost into something that we can defend, that we can fight from, and that we can survive in."

One of Geressy's platoon leaders, Lieutenant James Weber, would later write that the first sergeant, through experience and intuition, "knew the enemy would soon be testing us. He made the demanding but necessary directive for the company to continuously improve our force protection measures for forty-eight hours. In my opinion, the standard to which (1st Sergeant) Geressy required the defensive position to be improved would not have been reached had he not been present. And it was this standard that proved crucial in the defense of the COP, especially in the initial moments of the attack."

I didn't know the enemy situation, but I had been in the Army for a while by then and had been in combat several times. I had a feeling that we were going to be attacked, and soon. I came up with a plan right away and put everyone to work. We quickly built bunkers on the roof for our machine gun teams. We took a lot of what was called "pope glass," which is the type of bulletproof glass that was used for armored Humvees. There was a lot of that stuff lying around on

Falcon, so we took it for better use than having it sit all over the FOB. We also took a bunch of extra plywood, and sand bags. I coordinated with our headquarters company to sign for several extra M2 .50-caliber machine guns and two MK-19 automatic grenade launchers for the fighting positions on our roof. These extra weapons systems meant that all sectors of fire would be covered by heavy weapons for primary positions, alternate positions, and supplementary positions.

The soldiers were busy nonstop. It was more than one hundred degrees and they were out there filling sandbags, carrying them to the roof, building bunkers out of plywood and two-by-fours, cutting frames to hold the pope glass. We also erected a 60mm mortar position to be fired from the roof along with a storage area for its ammunition. We did a lot of the work at night so nobody could see what we were doing. Then we put up a bunch of camouflaged netting that screened the entire post. With those nets up, the enemy couldn't see our fighting positions. They couldn't see us moving around on the roof, either. Looking back, that camo netting probably saved more lives than anything else because it caused the enemy to shoot too high. The enemy fired at that net so much that during the attack some of the netting caught fire.

I remember thinking back to having the honor of meeting Medal of Honor recipient Lieutenant General Robert F. Foley. He served in Vietnam in the 25th Infantry Division, and he came to speak to our unit while I was stationed in Hawaii shortly after 9/11. He shared his experience as a company commander and about how to build a defense. He said, "Deception, deception, deception." It stuck in my head for years. So one of my ideas was to build much of the defense during the night to confuse the enemy. If they did attack us during the day, our positions would be different from what they saw previously. That's exactly what happened, too. They didn't see our new positions that we'd set up during the night.

The enemy did not realize the amount of improvements to the outpost's fighting positions that were made by the soldiers of Eagle Company. Looking back it was amazing what they accomplished in the forty-eight hours prior to the enemy attack. The insurgents thought we wouldn't be ready; they could not have been more wrong. Besides the position improvements, the deception that we capitalized on was essential. It worked for Lieutenant General Foley in Vietnam, and it worked for Eagle Company in Baghdad.

The soldiers worked feverishly on that first day and through the night, even wearing their night vision goggles so they could set up the camouflage netting. By our second day at Blackfoot, the defense was nearly 100 percent complete. They worked day and night with no complaints, although I don't think they were too happy about it. By the time the defensive positions were complete, the entire company was probably ready to catapult me from the rooftop and into that neighborhood. The soldiers' hard work had paid off and the last machine gun was in position by the morning of September 4. A few hours later, around 3 p.m., the battle started.

Blackfoot was attacked on your second day there?

Yes, it was. Our company had started to move to the outpost around September 1, and we had all three rifle platoons in place but were missing a large part of the headquarters and the mobile gun system platoon. You see, the enemy had figured out when new units were rotating into Iraq, and these units were the most vulnerable while they were learning about their new sectors. The unit we replaced introduced us to some of the local Iraqis, so word quickly spread that the new guys were moving in. We later learned through interviews that al-Qaeda had been planning to conduct a major assault on the American soldiers in that sector, and since we were new to the outpost they would capitalize on the situation. We also would learn that there were about 150–200 fighters in their assault.

Al-Qaeda's plan started on the previous night, though, with a little deception of their own. The night before, when Eagle Company soldiers were setting up the defenses, we got a call from the squadron's TOC, or tactical operations center, on Falcon. They said they had received intelligence about a vehicle-born improvised explosive device, a VBIED, which is basically a car full of bombs. These VBIEDs were extremely dangerous and would often create mass casualty situations. It was supposed to be in our sector, about 3–4 kilometers from Blackfoot, so the TOC wanted me to send Eagle Company soldiers to investigate.

My first question was, "Where are you getting this report from?" The squadron TOC said they received it from a tip line that the unit we replaced had established. That tip line was simply a telephone number with a voicemail box. The unit had advertised the number around Baghdad asking people to call with information about the insurgency, warnings about attacks, or where we could find wanted terrorists. But it's just an anonymous tip line. You don't know who you're getting the information from. It could be from anyone and for any purpose, good or bad.

So I asked the TOC, "You don't know where this report is coming from, and you want me to send a platoon, that's only been here two days, out into the night looking for a car bomb that might be there, or might not?" I thought for a moment about how we didn't yet know much about the enemy in the area, and then said, "Sir, I don't think that's a good idea." I didn't want to come off as insubordinate, but I just couldn't agree with the idea based off of our current company situation and the lack of knowledge about where the report came from. The battle captain at the TOC accepted that as an answer … for the time being. They kept coming back for the next few hours asking me to send the platoon. Eventually, when daybreak came, I got a direct order from the squadron TOC to send a platoon to investigate the reported car bomb. There was no way I could push back at that point.

I knew we had to do the mission, but my first priority was always the safety of my soldiers. So I assembled all of the platoon leaders and platoon sergeants, along with our physician's assistant, to brief the plan. The unit we replaced rarely went out during the day, so doing this in late afternoon would help because we would catch the enemy by surprise. The platoon I selected for the patrol was Eagle Company's 3rd Platoon, which was led by 1st Lieutenant Christopher Turner and Sergeant 1st Class Jeremy Hare.

The plan was for 3rd Platoon to move dismounted, which means on foot, from the outpost through the alleyways and streets to the suspected VBIED location. We kept their vehicles at the outpost with their crews ready to go, with engines running, in case they needed to be reinforced or for casualty evacuation. I also gave the same instruction to 1st Platoon, which was led by 1st Lieutenant Fernando Pelayo and Sergeant 1st Class Raymond Bittinger. Their task was to be loaded up with the entire platoon in case we needed to reinforce 3rd Platoon by maneuvering on the enemy or assist with casualty evacuation. I also requested Apache attack helicopters to provide over-watch during 3rd Platoon's movement through the neighborhood.

At 3:30 p.m. our Apache helicopters checked in, and then 3rd Platoon left Blackfoot dismounted. For better concealment, 1st Lieutenant Turner and his platoon crept through the alleyways rather than walking into the area on the normal road. If the enemies were there, they wouldn't expect to see our guys coming dismounted along that route. Meanwhile, back at the outpost, 1st Platoon was set as a Quick Reaction Force, a QRF, ready to ride out and provide backup should it be needed, and 2nd Platoon, led by 1st Lieutenant James Weber and Staff Sergeant Brian Glynn, was on the roof of the outpost providing security.

Now here's something interesting that we didn't know at that time: months later, we learned through intelligence gathering that al-Qaeda's initial plan was to use that tip line to lure one of our platoons to that suspected car bomb site during the previous night, then ambush and kill them all. We didn't take the bait, at least not during the night. After al-Qaeda's plan failed, the enemy leadership gathered in a mosque to decide what their next step should be because they still wanted to attack the newly arrived unit at Blackfoot. Coincidentally, the VBIED location was in the vicinity of that mosque where the meeting was being held. We were eventually told the enemy leadership went into a panic because they thought 3rd Platoon knew their location and was about to attack. So the enemy called all of their fighters together and had them mass for an immediate attack on 3rd

Platoon at the location of the VBIED, and then to attack COP Blackfoot after that.

It was incredibly hot, close to 120 degrees, and 3rd Platoon called and said that a couple of its soldiers were having a hard time with the heat and had gotten dehydrated. That can happen so fast over there. Also, as they were making their way to the area, they ducked into a building and found a whole bunch of bomb-making materials and a bunch of AK-47s with suppressors. The AK-47s also had a stamp on the rifle which identified that it came from the Iranian Revolutionary Guard al-Quds Force—the worst guys over there. More bad news.

Anyway, the platoon reached the suspected VBIED—sort of a bus—and a local Iraqi man told them that it had been parked there for months. When 1st Lieutenant Turner called me on the radio and told me that, I was convinced the whole "tip line" thing was a trap. I told the lieutenant that I was sending his platoon's Strykers to his location immediately, and for him to load the dismounted soldiers from his platoon back on the vehicles and return back to Blackfoot. But as soon as the Strykers departed the outpost, 3rd Platoon began taking small arms fire from enemy positions around the neighborhood. There were insurgents firing from rooftops and from windows. It was sporadic at first, but little by little the gunfire became more intense.

The Strykers got to the dismounted platoon within minutes. As the vehicles linked up with the platoon, they engaged and killed three al-Qaeda fighters who were firing at them from the rooftops. We were still very new to the area and didn't know what the enemy situation was, so instead of having 3rd Platoon maneuver on the enemy and engage, I decided to bring them back to Blackfoot.

As soon as the platoon pulled back into Blackfoot, I said to 1st Lieutenant Turner, "Sir, as soon as you can, get your guys around the map table in the command post and let's do a quick after-action report." I wanted to do this fast so we could get a better picture of the enemy actions. I then went to check on the soldiers who had trouble with the heat while on the mission. Captain Blanks and his medics had already put IVs into their arms and were cooling them down and getting them rehydrated. After seeing that they were all right, I went back to the command post to talk with 3rd Platoon's leadership.

I was the last one into the command post, and just when the platoon leader started to explain what happened, we heard the unmistakable sound of two or three rocket-propelled grenades zip right over the top of our outpost and detonate on the other side of the wall. Then we heard several other explosions, and then machine gunfire all around our building.

We had rehearsed our battle drills in case the outpost was attacked. All the soldiers knew their job—what they would do and where they would go in case this happened. Very briefly, I told 1st Lieutenant Turner and Sergeant First Class Hare to take their soldiers and reinforce 2nd Platoon, which was on the roof. At this point I had two platoons moving on to the roof manning the fighting positions. I then directed 1st Lieutenant Pelayo, with his soldiers from 1st Platoon, who were already loaded in their vehicles, to stay ready to go as the quick reaction force, a QRF. This gave me a quick option if I needed to decide whether to send them outside to maneuver on the enemy or keep them at the outpost ready to go for casualty evacuation.

Everything happened very quickly, in what seemed like two seconds. Everybody started executing and moving. At the time I was just wearing my uniform with only my pistol. My body armor, helmet and rifle were back in my room. I started running back there in what seemed like slow motion. I finally got to my room, threw on my kit—armor, helmet—grabbed my M4 and started running up the stairs.

Blackfoot was getting slammed. Initially, it was a shock to my company that this was happening so soon after our arrival. The award recommendation for that day says that I "calmly" did this and "calmly" did that. But I don't remember being calm about anything. I hit that first step in the stairwell and heard the guys on the roof screaming, "Medic! Medic!" When guys scream that word, you never know what you're going to find on the other end. The sound of that call for help, the panic in their voices … it's hard to describe, but I'll never forget those calls.

There were explosions impacting all around the outpost, and you could hear and see the impact of bullets from what I thought were several machine guns firing into our position. As I ran up the stairs I heard the enemy let loose with several more RPGs and mortar rounds. We had put sandbags in the windows of the stairwell, and as I climbed the stairs to the roof I heard bullets hitting the sandbags. I could hear bullets hitting the wall, as well. It felt like the whole city was firing at us.

When I reached the roof my first action was to identify the wounded and get them out of there. The previous day, I had the soldiers pre-position litters, extra first-aid kits, and ammunition in each doorway of the stairwell on the rooftop. All this was in place before the enemy initiated the attack.

The amount of incoming fire was massive, and I was actually surprised that we weren't all being shot off the roof. I saw Specialist Ryan Holly lying there, screaming and bleeding out very badly. He was in bad shape and was going into shock. The bullet missed his vest, entered his chest, severed his sciatic nerve, and then came out on the other side and stuck into the side of an M203 High Explosive Dual Purpose Grenade that he was carrying on his belt. Luckily, it didn't explode.

As we all moved to help Holly, one of our soldiers, Specialist Mike Foster of New Jersey, did a heroic thing. He was about twenty-one years old, and that was his first time in combat. He was a machine gunner inside one of the bunkers on the roof. As a gunner, Foster had a specific sector of fire that he was responsible for, but he saw all of us trying to evacuate Holley and how we were taking heavy fire. Now, nobody told Foster to do what he did next. He could have just stayed in his bunker, where it was much safer than being out in the open on the roof, but he saw we needed some help, so Foster took his machine gun off its tripod, came out of the bunker, positioned himself on the other wall and started engaging the enemy. This provided extremely valuable covering fire. Foster was awarded an Army Commendation Medal with Valor for his actions, and that's just one example of what the soldiers of Eagle Company did that day.

The next couple of minutes were a blur. I yelled for one of the soldiers to get the litter from where it was pre-positioned, we took Holly's grenade belt off, loaded him onto the litter, and then I yelled for someone to help me carry it. Specialist Tamim Fares, one of our company snipers who had run up immediately when Holly was wounded, grabbed the other end of the litter and we started carrying him down the stairwell as delicately as we could.

We were under constant fire from what seemed like every direction, but none of my soldiers cared about that. They just wanted to help get Holly to safety. About halfway down, Holly started screaming really bad again. It's hard to communicate in a situation like that, between the noise of all the explosions and gunfire and the screaming. Things are moving so fast, one hundred miles an hour, but then a moment comes along as if it were in slow motion. There in that stairwell, everything slowed down. I looked down at Holly and said, "Hey, man, don't worry. You're going to be all right. You're going to be all right." He calmed down a little. Our physician's assistant, Captain Benjamin Blanks, and his medics then met us halfway down the stairs. They took over from there and started carrying Holly away to the aid station.

Now that we had taken care of Holly, the next thing I needed to do was identify the enemy positions. I turned around and ran back to the roof. After a quick look around, I determined that we were being attacked from three different directions. We needed to gain fire superiority quickly, and I knew that we needed to attack their heavy weapons first, starting with their machine guns. I then went from position to position on the roof, directing the fires of our own machine gunners.

Earlier in the day I had coordinated for Apache attack helicopters to provide over-watch for the platoon that investigated the VBIED. They were still around, so I began coordinating with the pilots for a close-combat attack fire mission. My first target was a machine gun position firing from a four-storied building that was directly to the north of our outpost. Initially, we tried to show the pilots where we wanted them to shoot by firing tracer rounds from our machine guns into the target. But there was already so much fire going back and forth that the pilots couldn't identify the enemy using that method. So I told one of the M203 gunners from 2nd Platoon to fire two or three smoke grenades at the enemy location. He did, and it worked. The Apache pilots confirmed the smoke from the M203 that marked the enemy position, and I then cleared them hot to engage the enemy position using a hellfire missile.

The Apache fired the hellfire missile at the target. I watched it sail, and it was going, going, going ... and then right before impact it broke left and blew up a random house. I remember thinking, "Oh, Shit! I hope there were no civilians in that house." Thankfully the pilot came back on the radio and said, "Eagle Seven," (which was my radio call-sign), "That wasn't your fault. It was a missile malfunction." I then cleared the Apache for a second attempt on the enemy gun position. I think they did two strikes on that building, and those missile strikes silenced the enemy machine gun. I went through the same process clearing fires for the Apaches on several other enemy machine gun positions.

I cannot say enough about those helicopter pilots. They saved many lives, not only on that day, but also throughout our deployment. They were always ready and willing, never fearing to fly low or go on multiple gun runs in support of our soldiers on the ground. They were phenomenal, and I hope they know how much we appreciated their efforts.

In between the missile strikes, I was running up and down the stairs to the building's small command post to radio situation reports to Lieutenant Colonel Reineke, and at some time during the fight the colonel wanted to send the squadron's quick reaction force to reinforce Blackfoot. I paused to think about that. You see, during my second deployment in 2005–2006, the insurgents were notorious for ambushing the QRF, killing many American soldiers in the process. The enemy knew we'd always send help, so they would incorporate that into their plans. After they caused casualties during an initial attack, they would then just wait for a while and then attack the incoming platoon with roadside bombs and small arms fire.

I wasn't sure if we were going to hold the outpost, but I still didn't want to risk the lives of the soldiers on that QRF platoon. To me, the risk of sending them into the battle was just too great until we absolutely needed them, and I didn't know yet. Besides, this was Eagle Company's fight—we started it and we would finish it. So I recommended to keep the QRF ready, but not to send it in just yet. The colonel said he'd keep them loaded and ready to go, and that all I needed to do was say the word and they'd be released.

I got off of the radio and ran down to check on Holly. There were bullets ricocheting everywhere inside the building while Captain Blanks and the medics were working on him. You know it is bad when the medics in the aid station are working on the wounded wearing full kit—body armor and helmets. It was a bad scene. I asked the doc, "Hey, how is Holly?" Captain Blanks looked up and said, "He needs to get out of here ... and soon."

I then ran back up the stairwell to the roof to get a better assessment of the situation. By that time the fighting had been going on for over three hours, and while there was still a good bit of fighting, things seemed to have calmed a little. I felt that we had enough control of the situation, so I decided to have 1st Lieutenant Pelayo and his platoon evacuate Holly to the 28th CSH [*combat support hospital*]. They'd have to get through the neighborhood and then cross about fifteen kilometers of Baghdad. I told them to drop Holly off, leave one guy with him, and then haul ass back to the outpost. They immediately loaded Holly into a Stryker, and then I ran back to the roof. As soon as they pulled out of Blackfoot's gate it seemed like the whole city started shooting at us again. We were firing everything we had to suppress the enemy so Pelayo's platoon could get on the road: .50-cals, Mark 19s, M240s, and our snipers were engaging targets, as well.

The battle continued for what seemed like several more hours. There would be a lull in gunfire for a little while and then, without warning, everything would erupt again. We had most of the ammunition on the first floor, so between radioing situation reports back to our squadron's tactical operations center, asking for updates on Holly, directing gunfire, and checking weapons, I also made several trips running ammunition to the machine guns on the roof.

There was another lull in gunfire as we transitioned from daylight to nighttime operations. We started putting on our night vision goggles and checking our lasers, making sure everything was ready to fight in the night. It had been balls-to-the-wall fighting for several hours, and during that lull a few guys decided to take a knee on the roof and rest for a moment. One of them was Staff Sergeant Darrell Card, who is one of the best noncommissioned officers in the Army and a true hero. He looked and me and said, "Man, did that just happen?" I said, "Yeah, it did ... and I don't think it's over yet." Probably ten seconds after that, the gunfire erupted again.

We learned later that al-Qaeda was massing again and readying itself for another assault. They had planned an initial attack and then a supporting attack. They were even bringing in reinforcements and conducting ammunition resupply runs during the battle. That's how organized they were. Their supporting attack ended, though, after we had the Apache pilots conduct several more runs using their 30mm chain gun. That killed between twenty, twenty-five fighters. 1st Platoon then made it back to Blackfoot after evacing Holly, and the Apaches remained on-station. I don't remember how many different sets of Apaches supported us that day—they would go refuel and rearm, and then come back—but they saved many American lives.

There was sporadic gunfire back and forth throughout the night, and I had our soldiers firing parachute flares into the night sky over the neighborhood in order to remind the enemy, "Hey! We're still here! We're still ready to go!" But by midnight all had gone quiet, which was a good thing because we had shot damn near all of our machine gun ammunition and our MK-19 ammunition. We actually had to cut into the emergency ammunition pallet. A few hours later into the night we were resupplied, and that marked the end of the battle.

Soldiers fire at insurgents from atop Combat Outpost Blackfoot,
September 2007 (*E. Geressy*)

A few days later we heard that the surviving al-Qaeda fighters held a brief show-of-force in our sector by marching through a neighborhood street with their weapons. They were trying to look strong and demonstrate that they were still there, that they were still strong. But we had decimated them in the battle for Blackfoot. They were nowhere near the fighting force that they were the day before that attack, and even though there would still be many battles in the months ahead, that fight was the beginning of the end for al-Qaeda in our sector. The Iraqi civilians in the neighborhood knew al-Qaeda was finished as well. That was a decisive point for the people.

Whatever happened to Ryan Holly?

Holly survived. He was bleeding to death at our outpost, but Captain Blanks and our medics saved his life. Holly eventually got shipped back to the States, and he had a hard time recovering. Remember, the bullet cut through him, hitting his nerve before getting stuck in a grenade. A few days after the battle we had the explosive ordnance disposal team, the EOD, take the bullet out of that grenade. We then had it mounted on a plaque and sent it home to his family.

I'll never forget that day. Al-Qaeda's attack on Combat Outpost Blackfoot was massive and their objective was to inflict significant causalities to derail the positive assessment of the surge, but we didn't lose a single soldier during the battle. Looking back, all of the preparation and training that we did in Germany and Kuwait, and the performance of the soldiers during the fight, saved all of our lives.

We trained hard, and I pushed them all very hard. In the end, they all wanted to do good ... and they did. Every one of those soldiers is a hero. They went above and beyond what they were required to do. The soldiers took care of each other during an incredibly difficult time, and the lieutenants and non-commissioned officers really led their men. I am very proud of them, and our country should be very proud of them, too.

After that battle on the 4th of September, the fighting was door-to-door, house-to-house, and block-to-block until we slowly retook the entire neighborhood from the enemy. The civilians eventually came back. They opened up stores and started resuming somewhat normal lives. We even started inviting the Iraqis to barbecues on our outpost, and they would invite us for tea and dinner during our patrols. In the short time we were there, it went from one extreme to the other.

Our regiment would pay a heavy price in wounded and killed during that deployment. During the fighting to come, the regiment eventually had eighteen killed in action and seventy-five more wounded. Eagle Company had close to twenty-five wounded, and one killed in action. His name was Specialist Avealalo Milo, and he was about twenty-three years old. There were a lot of sacrifices over there. Too many ... a lot more than people realize. These soldiers really are our nation's best and brightest; the United States can be proud of how their sons and daughters conducted themselves during some very difficult days. These soldiers demonstrated valor, honor, and selflessness on a daily basis. It was an honor for me to serve and fight alongside those heroes. I think about them and the sacrifices they made every day.

Months later, during some local interviews conducted by 1st Lieutenant Patrick Rice, the Eagle Company fire support officer, an Iraqi who was part of the attack told us we had fought off an enemy force of up to two hundred al-Qaeda fighters, and he knew firsthand that the al-Qaeda commander had wanted to influence the media by staging a dynamic attack shortly before General Petraeus testified to Congress about the results of the surge. They knew what was at stake—the U.S. Army was either going to stay or retreat based on what the American people thought of what Petraeus said. A massive attack, wiping out an entire platoon or company on the eve of his testimony, would have been a great win for al-Qaeda. They wanted those headlines.

We learned that they had done all kinds of reconnaissance on Blackfoot prior to our arrival, and they thought that a new unit like ours wouldn't be ready so soon. They thought they would catch us unprepared, and kill many of our soldiers. They wanted to wipe us out. That was their plan, but their attack on Blackfoot, and their plan to derail the surge, failed.

The entire squadron was recognized with the Valorous Unit Award for their actions in September and October of 2007, and Geressy was eventually promoted to sergeant major. Many officers in the unit would later write about Geressy in a packet seeking official recognition for their first sergeant. 1st Lieutenant Turner, one of the new officers in the company, wrote fondly of Geressy, reflecting that, "As a young platoon leader under fire for the first time, Sergeant Major Geressy's leadership, poise, and heroism set the example for me to lead my platoon through the grueling firefight." Captain Blanks, the physician's assistant, wrote that during his more than seventeen-year career in the military, "I have never served with a more courageous and capable leader than Sergeant Major Geressy."

For his leadership during the battle for Blackfoot, Eric Geressy's company commander recommended him for the Silver Star. Lieutenant General Ray Odierno, then-commander of Multi-National Corps-Iraq, was so impressed by the action that he upgraded the award to the Distinguished Service Cross (DSC), second only to the Medal of Honor. Surprisingly, the general's upgrade somehow went unnoticed during the process and the Silver Star was presented to Geressy in a ceremony at Fort Bliss, Texas, in September of 2008. Geressy's grandfather, a veteran of World War II, was at the ceremony and in a touching moment pinned the medal on his grandson's uniform.

In the years since, Geressy's former commanders and soldiers began advocating that he should receive the DSC after all, and in 2010 an official recommendation to upgrade the award was sent to the Secretary of the Army. On the form recommending the upgrade, General Petraeus personally wrote, "Truly heroic and deserving of DSC," and General Odierno, who would eventually become the Army's chief of staff—its top-ranking officer—also wrote that Geressy's "actions warrant the DSC. I truly believe he is much deserved of this award."

The board reviewing the request ultimately disagreed with the generals, and the upgrade to the DSC was denied. During an appeal process, it was noted that one of the board's members said that Geressy's actions were "business as usual for a sergeant in combat" and that there was not "a lot of loss of life and not a lot of saving of life" during the battle. Another board member who voted against the upgrade said Geressy "did not fire a weapon or engage personally with an enemy soldier" while leading the defense of COP Blackfoot.

Veterans of the battle—those whose lives were saved by the actions of their first sergeant—were left stunned by the board's rationale. The inexplicable decision could still be reversed, however. In early 2016, the Pentagon began reviewing more than 1,100 awards for valor that were issued since September 11, 2001, with an eye to upgrade those who deserve higher recognition. The sweeping review comes after many years of protests from Congress and veterans groups who claimed the process left many warfighters and their families without the recognition they rightly deserve. "It's a systemic problem," said U.S. Rep. Duncan Hunter, a California Republican and Marine veteran. "I'm glad they're finally getting around to fixing it. This is military bureaucracy at its worst."

Meanwhile, Geressy remains loved and admired by the men who fought alongside him at COP Blackfoot. "With absolutely no doubt," 1st Lieutenant James Weber wrote, "Geressy's crucial preparations, flawless leadership, and extraordinary heroism saved the lives of an untold number of American soldiers that day."

Eric Geressy with his grandfather, World War II veteran Mitchell Rech, and Colonel Mike Steele, after being presented with the Silver Star, Fort Bliss, Texas, September 2008 (*U.S. Army*)

CHAPTER THIRTEEN

"IT WAS KILL OR BE KILLED."

JARION HALBISEN-GIBBS
DISTINGUISHED SERVICE CROSS, *SAMARRA*

O n Tuesday, September 4, 2007—the same day that Eric Geressy
and the men of the Second Dragoons were defending Combat
Outpost Blackfoot from the massive attack— their commanding
general, David Petraeus, arrived in Washington, D.C., and began preparing
to deliver his make-or-break testimony to Congress about the surge.

Thom Shanker of the New York Times captured the suspense that was building in the capital: "The testimony about the status of Iraq that General Petraeus
will deliver to Congress beginning Monday has become the most anticipated by
an Army officer since April 29, 1967, when, under President Johnson, Gen.
William C. Westmoreland traveled from Vietnam to address a joint meeting of
Congress at a time of deep public doubts about a faraway war."

General Petraeus and his top staff aides spent the next few days writing, and then re-writing, the draft of his remarks several times. The
remarks not only had to be accurate and reasonably brief, they had to be
remarkably compelling—a few minutes of an individual's testimony
would be weighed against the feelings left after four and a half years of a
bloody war. Many were convinced that the war was lost, while a few
others remained open to new ideas, and were hopeful that the general
would be both credible and convincing. The speech was finally finished on
Sunday evening, September 9th, only hours before General Petraeus and
Ambassador Crocker were due before a joint hearing of the House armed
services and foreign affairs committees.

As those in Washington slept on the eve of the general's testimony, on the other side of the world in Iraq, Jarion Halbisen-Gibbs and his team of Green Berets were in a fierce nighttime gunfight after hunting down one of al-Qaeda in Iraq's senior leaders. This was the Special Forces soldier's second tour in Iraq, and he was implementing the many lessons he had learned from his first. For someone who once thought that he missed the war, Halbisen-Gibbs was about to get more than enough. "We'd gotten in some pretty good firefights back then, and my team gained some very valuable experience, but that first deployment was nothing compared to my second," he told me.

JARION HALBISEN-GIBBS: I was our Special Forces team's senior weapons sergeant when we deployed to Samarra in 2007, and I was about twenty-six years old. I had gone to a couple of great schools since my first deployment in 2005—the combat dive school and the Special Operations Target Interdiction Course, known as the Special Forces Sniper Course—so I felt much more prepared.

I was a staff sergeant, still with Operational Detachment Alpha 083, and we deployed back to Iraq in March of 2007, right at the beginning of the surge. Samarra was a busy place, to say the least. It was a hotbed for insurgent activity because the surge pushed many fighters out of Baghdad and into smaller cities like Samarra. It was a very busy, very violent time in the war. The American presence in Samarra was very small at that time. In our areas there was just one Operational Detachment Alpha, which was us, and a company-plus of 82nd Airborne guys, which was a little more than two hundred paratroopers.

Samarra is a pretty big city. There's a lot of farming and agriculture around that area too, and the Tigris River moves right down the western side of the city. It wasn't nearly as urban as Baghdad, though. I wouldn't necessarily say it was run-down, but you could definitely see the evidence of battles, both recently and long ago. Between all the insurgent fighting, and then the Americans coming in, Samarra was the very definition of a war-torn city.

There's also a great deal of historical significance in that town, and it has a couple of very important religious features, so it's an incredibly symbolic position to either defend or capture. It's home to the Great Mosque of Samarra, which is also known as the Spiral Minaret. It's a cool thing to see, especially to be able to look out from the top of your team house and see it every morning. The Al-Askari Shrine, the mosque with the golden dome, is also there. It's one of the most sacred locations for followers of Shiite Islam. While I was there the shrine was blown up for the second time. It was a Shiite mosque in a Sunni town, so automatically you're going to have some sectarian fighting.

Before the shrine was attacked for the second time, the city's local police force had been hit with a very large VBIED [*vehicle-borne improvised explosive device*]. It was essentially a van fully packed with explosives, and it detonated about six hundred meters away from our team's house. That may seem a safe distance, but the blast was so large it shook our house and the shockwave felt like an RPG had struck the outside wall of my room.

The blast completely leveled many of the buildings in the area. In all of my experiences, the aftermath of that attack was the worst I have ever personally seen. It was a mass casualty event. Body parts were strewn among the rubble, and everything was covered in a thin layer of gray ash from the remnants of exploded concrete. This made the red blood stand out in a grotesque fashion; I'll never forget how its bright color contrasted against the ashy coating as it flowed from the dead. A water line was broken during the blast, as well, and the street was swamped in standing water. It mixed with the blood, and the reddish waters flowed down the street and away from the blast. It was horrible.

That explosion broke the back of the local police force. Many of them died in the blast, and the ones who remained were totally shellshocked. Cops were walking around dazed, covered with ash and their clothing shredded. The town became lawless after that because most of the police force was either hurt or they'd gone back to their families.

What was your mission there?

We were there to do what Green Berets do: make friends, learn who the good guys were, learn who the bad guys were, and then take care of business [*laughter*]. I need to learn how to phrase that in a nicer manner, but it's essentially what we do. Anyway, at that time during the war, we weren't going unilateral, meaning conducting raids without partner nation forces. So if we were actively hunting down bad guys, we had to have an Iraqi partner force with us. This allowed us to both train the Iraqis while putting an Iraqi face on our operations.

Jarion Halbisen-Gibbs after a raid on the outskirts of Samarra, spring 2007 (*J. Halbisen-Gibbs*)

**Jarion Halbisen-Gibbs pulls security near the Special Forces team house,
Samarra, summer 2007 (*J. Halbisen-Gibbs*)**

One of the aspects of Special Forces that worked very well to our advantage is that we're culturally aware. Sometimes the picture gets painted that Americans simply go out with all of this firepower—bombers, helicopters, tanks and all that—and that we just kick doors down and blow stuff up. In certain cases, that is true. But in many cases, it's not. If you want to really affect an area, you have to know who lives there. You have to know who controls the neighborhood, who the elders are, who the decision-makers are ... who has the real power. If you can figure that out, especially in a tribal nation like Iraq, you'll want to make friends with them, and if they're on your side, you want to empower them.

Remember, back in 2007, our military as a whole was just starting to figure out how to do tribal engagement. It was not a tactic that was taught in training courses or in military colleges. There wasn't a manual on it. But our Special Forces teams were doing some really successful tribal engagement in our area at that time. Through empowering one of these local tribes, they not only rolled over on the bad elements within their network, but they also identified the local troublemakers and the foreign fighters. It's easy for them to identify what's happening because it's their own backyard. It's not ours. Working by, with, and through the Iraqis is how we were able to be so successful, as opposed to just looking at a map and trying to figure it out.

Working with a culture that is so different from ours, especially with a language barrier, may seem insurmountable, but it's absolutely not. Being culturally aware, paying attention to your surroundings, and simply being respectful to the local population pays dividends in how people are going to perceive you. That's not just in the military. That's in everyday life. If you go to another country, you don't want to be the ugly American.

That's what tribal engagement comes down to. It's a working relationship, and you need to figure out what "right" looks like—together. What right looks like in America, or for the American military, isn't necessarily what it's going to look like for Iraqis or any other foreign military or nation. At the end of the day, what "right" looks like is simply what's successful for them. They might have different physical training necessities that the Americans wouldn't like or wouldn't see the same way.

You have to gain their respect, and once they respect you, they're willing to fight with you. You can gain a lot of that respect by living with them, as well. A lot of military units would train with their Iraqi counterparts during the day and then leave and go back to their base at night. The Special Forces wouldn't. We had our guys based and housed right next to us, literally. It was more than a working relationship; it was a living relationship. I could get up in the morning, walk right next door and grab one of my Iraqi squad leaders and say, "Hey, man. Let's go out for a run. Let's go lift some weights. Let's get breakfast together." That builds camaraderie, and when you make friends with people it just makes everything else easier. Special Forces guys had been doing this since the Vietnam era, so it was quite normal for us to continue doing so.

Many of the "jundis"—Arabic for soldier— were notoriously unreliable when placed with units that didn't know how to properly train or command them. Chris Kyle wrote in his autobiography that the Iraqi soldiers he encountered were often "pathetic" and sometimes even placed the mission at risk. "As fighters went, they sucked," Kyle wrote. "The brightest Iraqis, it seemed, were usually insurgents fighting against us. I guess most of our jundis had their hearts in the right place, but as far as proficient military fighting went, let's just say they were incompetent, if not outright dangerous." Thankfully the Green Berets were experts at training and fielding foreign fighters, and they often had the best of the Iraqis to work with.

We'd do firing range training or tactics training with the Iraqis during the day, and then in the afternoons when it'd get really hot we'd find a nice cool room, something with a white board on the wall, and just talk about and teach tactics. Then after dinner, after nightfall, we'd gear up and conduct missions. Now, if there was one downfall in the process it was this: we didn't have a chance to just train the guys. It was training and serious operations happening simultaneously.

Tell me about the mission on September 10th.

It was named Operation Chromium, a raid to capture Abu Obadiah, who was al-Qaeda in Iraq's Minister of Defense for the entire Salah ad Din Province. He was a time-sensitive target that we had waiting on a trigger. That day, the trigger went off so we got ready to execute the mission.

Trigger?

That's something that we're waiting to happen before we conduct a mission. It could be anything, any piece of intelligence that tells us the target is ready to be taken. Maybe it's a certain car pulling into the compound, for instance, or maybe a multitude or combination of other things. But once that trigger goes off, the team has to be ready and fast enough to execute on that target before the person, or whatever you're going after, can move.

We were very lucky up to that point. We'd been conducting a lot of missions and we were getting the guys we were after, along with a significant number of weapons caches. We had been working our way up the chain hoping to target some leadership. That night, the trigger went off for Abu Obadiah. He was the number-two guy on our high-value target list.

What happened when the trigger went off?

We gathered our assault force, which was our Special Forces team and a section of the Iraqi police force's SWAT team, and drew up a quick plan on the whiteboard. We started out simple; here's our objective and here are our teams going in. Then it was all just details after that.

We broke up our Iraqis by colored chem-lights because they didn't have night vision. One team of Iraqis would be following the green chem-light, which was attached to my body armor, and the second team would follow a guy with a blue chem-light.

We had to do this because when you're conducting nighttime operations it's very hard to move if your counterpart doesn't have night vision. One way around that is to attach a chem-light to the team leader and tape it up so that only a very small portion shows. Then when we get off the helicopter and we start moving, all my Iraqis could very quickly find the green light and know to follow me. I would lead them to the objectives, to the breach points and, once we got to the house and started conducting close quarters battle, they could go "white lights on," which means they could use the flash-lights on their weapons.

We broke our task organization into Assault Force I and Assault Force II. I was on the second, with Captain Matthew Chaney, the team leader, and Sergeant 1st Class Michael Lindsay, our senior communications guy (*who was also in the 2005 engagement when Jar used the minigun to thwart an ambush*), and then six or seven Iraqis.

We were on the first Black Hawk inbound. The second team had three other Special Forces guys and a few more Iraqis. They were going to hit another portion of the objective. Then we had another helicopter that carried a few guys who were going to be a blocking force and provide supporting fire from afar. So in case the bad guys ran, they'd be able to stop them. We were supposed to have a fourth helicopter that night. It would have carried our K-9 guys, with the military working dogs. They do a great job and definitely helped us during the deployment, but as luck would have it their helicopter broke at the airfield, so the dogs and their handlers stayed at the team house.

Later that night the helicopters picked us up at the HLZ [*helicopter landing zone*] outside of our team house, and we flew towards the objective. It was a three-building compound on the rural outskirts of Samarra. Originally, we were planning on doing an offset infiltration. That means we were going to land in a field relatively far away from the compound so that the bad guys wouldn't hear us coming. Then we were going to walk to the objective, set up a support-by-fire line, and then send our two assault forces to hit different portions of the objective.

But as we started flying closer, the aircrew radios back and says, "Hey, there is standing water in the field so we can't land there. What do you want to do?" Talking to the team leader, the captain, we made the call right there on the spot. "We're going to have to change the plan. Put us down on the 'X,'" he said. That meant land right outside of the house. The crew was comfortable doing this, but it changed the dynamics completely. The two assault forces were going to be immediately in the fight, while the guys aboard the third helicopter instantly changed from a blocking and supporting fire force into an aerial quick-reaction force. They were going to stay in the air, and wait to see when and where they were needed the most.

It was a very dark night, only about 10 percent illumination, which means it was so dark that even our night vision was grainy. I and the other two Special Forces teammates had our night vision, infrared lasers on our rifles, and infrared floodlights. We could see what our laser is doing at night. Basically, where the dot went, that's where the bullet went. It makes it very easy to shoot, and the infrared floodlight illuminated the area around the laser sight, too.

As we came in, we were able to see what the buildings looked like, and see that there were no lights on. From what our reports had told us, it was supposed to be Abu Obadiah and about six of his bodyguards … but there ended up being far more people in those buildings than we thought. My bird was the first to touch down. We landed about thirty meters off of the first objective, which was a small building on the south side of the compound. As we came in, the helicopter flared and they landed right next to the first structure. We were running as soon as our boots hit the ground, and the helicopter quickly took off.

I made it to the structure, and our Iraqis came in behind me. We cleared the first building, which was just something for the farm's livestock. There was nothing in there. So I pushed our Iraqi guys back outside, which is when we started hearing fire coming from somewhere in the compound. The bad guys had obviously heard the helicopters and were getting ready for the fight.

When you're taking fire, you move toward the fire, not away from it. You must suppress it, and then take the fight to the enemy. Never give them time to react and move on you. So, after hearing the shots I had started moving my team towards the sound of gunfire. Then the second helicopter carrying Assault Force II came in. Because the farm was so dusty the landing blacked everything out, creating what's known as a "brownout." The dust rolled up and we couldn't see anything for a few seconds. That was when our Iraqis got disoriented and broke away from us. Even though it became very difficult to see, Matt, Mike, and I continued toward the objective.

As the dust cleared, we found ourselves right outside of the objective. It was a small, one-story L-shaped building. It had cinder block walls, a thatched roof of long reeds and grass, with chopped-down trees as the support beams. They stuffed mud between the cracks of structures like that. It had one large room, and then some smaller rooms that came off the end of the building forming basically the shape of an uppercase letter "L."

I looked at the structure and a guy with an AK-47 stepped out and started running down the side of the building away from us and towards where Assault Force II had landed. Mike was on my left and Matt was to my right, and I yelled back at my guys and said, "Hey, they've got AKs." As I said that, I put my rifle on fire, centered my laser pointer on the target, and shot him about ten times with M855 rounds before he hit the ground.

Now, the M855 round doesn't do a whole lot of damage. It has the ability of ice-picking somebody, if you will. It's a very small round. It's moving very, very fast. When it goes through a person, if it doesn't hit a bone or something like that, it just creates a very small hole. You can shoot somebody multiple times but it might take them a while to actually die. So, like I said, I shot him about ten times before he actually went down.

After that, everything got very hectic. Mike started shooting another guy at the far end of the L-shaped building. As the first guy I shot was falling, I was already stepping to the door of the building where he had exited only seconds before. There I saw another guy, who was standing in what we like to call the "fatal funnel." He was silhouetted in the doorway trying to load a magazine in his AK. He was probably in his twenties, a thin guy, wearing a loose pair of pants and a T-shirt. He didn't have a vest on. He was having a difficult time finding the magazine well of his AK, costing him valuable seconds. So I aimed a little lower than usual because I didn't want to hit his gun, pulled the trigger, and dumped about five rounds through his solar plexus. He fell backwards. That kill came only a second, maybe two seconds at most, after the first. It was all happening very fast.

At that moment I felt like I was doing what I was put here to do, being a warrior not only in word but also in deed. The seconds slowed and, once again, I felt like I could see every detail of the situation. Some people who have been in combat describe it to civilians as similar to being in a car crash, when the accident seems to play on slow motion, and your senses are heightened and you're noticing everything, including the passage of time, with a greater degree of clarity than usual. It's like that, but only partially. In a car crash, you're not in control. You're just along for the ride. In a firefight, with training, dedication, and experience, you can learn to operate in the chaos.

So as the second guy I shot fell backwards into the room, the door was somehow pushed closed behind him. The door was made out of tin, so it was definitely not going to stop bullets. As his fall shut the door, I could see people in the room, scurrying and running around, so I put another couple of rounds through the door. I got around a corner and Matt yelled, "Hey! Get a frag [*fragmentation grenade*]." I said, "Yep. Already got it." Mike got behind the captain, or maybe he got behind me, I'm not sure which. Anyway, we were stacked up outside ready to go in. I grabbed my frag and tossed it through a small opening in the door. Then I pulled back around the corner and waited for the longest five seconds of my life for the frag to go off, and then it detonated. [*Three insurgents were killed by the blast.*]

As we were pulled back waiting to go into that room, Assault Force II made it to us. Because of the brownout conditions and the ambient noise from the helicopter, they didn't realize that we were already in a firefight. When we pulled back and threw the frag in the room, they thought that we were using the flash-bangs to prepare for entry. With us being stacked up on that building, they bypassed us, leap-frogging us to the next structure. They went to the other side of the objective to clear the second building, which was exactly the right thing to do. I didn't know that because we were focused on that room, but I'm glad they did, because that took another piece of the objective off the plate.

After our frag went off in that room, I came around the corner and had to kick and shoulder my way in, pushing the door out of the way to get in. The second guy who I had shot, he had fallen back and his legs were in the way. As I came into the room, it was hard to see because it was dark and smoky because of the hand grenade explosion. Frags have a very specific smell, and a specific kind of smoke. It's really dry, and especially inside rooms like that, it's very black. Some other explosives will leave a much heavier residue, but a lot of them pretty much smell the same way. There was also dust and debris floating everywhere, of course, because when a grenade detonates in a small room like that, the overpressure blows everything into the air—dirt, loose paint, dust, just about anything small enough to become airborne.

We entered the room and immediately started engaging our corners. I centered my laser onto a guy in my corner and shot a couple of rounds into him. Matt and Mike were right behind me, and firing, too; I could feel the overpressure from their weapons since they were shooting right over my shoulder. Then somebody from the opposite corner of the room started shooting back, and I got hit in the thumb of my right hand.

It was a ricocheted bullet, and the impact was bad enough to break my right thumb, laterally fracturing it. When it hit, it pinned my thumb against the side of my rifle. A ricochet caught the tip of my right middle finger, too. Luckily, I was wearing gloves and I didn't know how bad it was. Later, at the hospital, they took off my glove and you could see right down to the bone on my thumb. To tell you the truth, getting shot in the thumb hurt worse than anything else that happened that night. It was seriously painful. You have a great deal of nerve endings in your fingers, and not so many in your gut, for instance. So while you hear in the movies that being "gut shot" is the worst pain imaginable—and trust me, it does hurt—getting shot in the finger is much more painful.

They hit Matt, too. The bullet went through one of his butt cheeks, bounced off his pelvis, and exited out the other side. It was the Forrest Gump million-dollar wound. They hit Mike with a couple of rounds, as well. We were getting shot up pretty bad. Then all of a sudden there was a flash of light, intensely green because I still had my night vision goggles on. Turns out, one of the enemy fighters had pulled the pin on a grenade and it had detonated inside the room. It's funny because I still don't remember feeling the impact. I still don't remember feeling overpressure or a blast. I was fighting. I was shooting, and then ... bam! There was this flash, and I was out. At the hospital later that morning I learned, in addition to being knocked out by the blast, I was also hit in the arm, on my left bicep. The scar is still there.

I came to on my knees near one corner of the room, on top of the second guy who I'd shot, who had fallen back through the doorway. He was still alive. I could feel him squirming around underneath me, and my knee was in his abdomen, right underneath his rib cage. I could feel where he was bleeding out all over my knee. I was still tossing the cobwebs around in my head trying to figure out what the hell was happening. It was at that point I realized a second hand grenade had gone off and I'd gotten blown up, and then thrown on top of this guy. I didn't know it, but Matt and Mike had been blown outside.

I had killed quite a few insurgents by this point in my career, but that one was different. He was underneath me, fighting for his last breath, and I was literally crushing the life out of him. I had to, because I didn't know what he was doing, or what he was trying to do. He already showed his willingness and intent to fight us, and I didn't know if he was going for his weapon or a grenade. I only knew that it was kill or be killed. So I kept the pressure, kept the force of weight on him while trying to scan the room for additional threats. That kill was pretty gruesome. It was intensely personal, with no barrier between him and me. It affected me pretty deeply.

Could you see?

Partially. When the second hand grenade detonated, it had blown out the left side of my dual tube night vision goggles. I'm lucky that I had dual tubes because without that gear to absorb the impact the left side of my face probably would have been blown off. Aside from that fact, even though the left side of the NVGs got blown out, the right side still worked, so I could see out of that.

Could you hear?

Yes. Luckily, I was wearing noise-canceling headphones with a wire that plugged into my radio. With them, I could hear radio chatter and what's normally heard in a close conversation, but anything louder than that—anything over about eighty decibels—the headphones cut off the sound. They're really good for protecting your hearing, and between having the headphones and night vision goggles on, and wearing body armor, I was pretty much okay after getting blown up by a grenade. The blast definitely knocked me unconscious for a second but I came right back and was still in the fight.

Looking back, they were trying to throw that second hand grenade at us. As luck would have it, we had the drop on them and we were faster. When I threw the grenade in the room and it went off, whoever was trying to throw a grenade out at us had it blown out of their hand.

When we enter a room or conduct CQB [*close quarters combat*], it's pretty freaking fast. The time it took us to come around the corner, get in that room and have our initial firefight, was the same amount of time it took for that second hand grenade's timer to go off [*about 3–5 seconds*], which blew me into the wall and then blew Mike and Matt back outside.

There were a lot of guys in that room. We later learned that they had tactical vests hanging on the walls with AK-47 mags and hand grenades. They had go-bags ready. They had RPG7s, PKMs, and AKs stacked in the corners, ready to grab and fight. These guys were pretty squared away. They were ready to go. We were just lucky enough that we got the drop on them, and that they weren't able to react. If they would have had even another thirty seconds to grab their guns and come out, they would have caught us wide out in the open without cover or concealment, and caught the birds, too. It would have been a bad night for us. Luckily, speed was our security.

It was a very hectic situation. But again, that's what we train for. You can't control chaos but you can certainly understand how to operate in chaos. That's why we do so much training, to keep familiar with situations and scenarios like that.

So, with the guy still underneath me I tried to look around the room. There was so much smoke and dust and haze that I really couldn't see much. I was looking for strobes, because our guys would have an infrared strobe on their kit or on their helmet so other assaulters or aircrew could easily identify them. I didn't know if Matt or Mike were still in the room, or if they were blown out, if they were okay or hurt. I didn't know. Honestly, that's a horrible position to be in—stuck in a room full of death and not know where your brothers are. I looked around, and I could hear guys dying. A lot of the bad dudes in that room just curled up, "O Allah, O Allah, O Allah." You know they're dying.

After multiple grenades had detonated in the room, and all of the gunfire, it was smoky and hard to see. As I looked around, all I heard was guys dying. I didn't see any strobe lights. Then, from the corner of the room somebody shot off a round. The bullet skipped off the floor, went through my right boot and cut into one of my middle toes. I could feel the bullet rattling around in my boot.

I was like, "Damn it. I just got shot in the thumb. Just got blown up. Now I'm lying in the corner, trying to hold this guy down and crush the life out of him. I don't know where my buddies are, and now some asshole shoots me in the toe?" Pardon my French. It was definitely time to get out of there and get Assault Force II to come back and mop up. [*After a few more moments, the insurgent underneath his knee stopped moving. He was dead.*]

That was a tough moment. On one hand, there were obviously still insurgents in that room capable of fighting, yet on the other hand, I didn't know where my brothers were. I didn't know they had been blown outside of the room, so I couldn't just open up with my M-4 and start blasting away because I might hit one of my own teammates. That still haunts me; I don't know if I made the right decision, but I couldn't just sit there wondering. We really had to finish the fight, or we'd all be done.

So I decided to leave the room. I jumped up and ran outside, and then I immediately saw that Matt and Mike were outside as well. Matt was taking cover behind the first guy that I shot, and shooting at another insurgent on the other side of the house. The dude presented himself out of this doorway as soon as I came out of the room. As I looked over and saw this guy, ready to shoot me. He had me dead to rights, but Matt took him down. Thank God that Matt was there.

Then I saw Mike, lying in the middle of the courtyard. He had taken the brunt of that grenade, and it blasted in his guts pretty badly. He had also taken a round a couple millimeters from his jugular. It's a hell of a scar now. He was so badly hurt that he couldn't even raise his rifle, but he still had enough strength, and willpower, to raise his pistol and shoot into the doorway of the L-shaped building. That was amazing; he was still in the fight.

I started moving towards Mike, with the idea of grabbing him by the kit—the type of harness that we wear—and bringing him out of the kill zone and safely behind a wall. But as I got about six feet away from Mike, another insurgent popped out of a doorway in the L-shaped building and got a lucky shot off with his pistol. I'm still not sure if he was shooting at Mike or me, but the bullet went straight through my radio battery, which was attached to my gear near the abdomen area, through my small intestines, through my right iliac crest of my hip, and then stopped in my "buns of steel," as I like to say.

Ouch! Hate to ask, but how did that feel?

Honestly, at the time I was so jacked up on adrenaline I didn't feel much. I felt the impact, sure, but the rest of the feeling was something akin to the jolt felt after touching an electric fence. Impact and pressure, but not pain. My thumb hurt worse, believe it or not. My thumb was throbbing, and I didn't know how much blood I was losing. All I knew for certain was I came out of that room completely furious, probably more pissed off than I've ever been in my life.

After you were shot, did you fall?

No. I was still standing and was in the middle of the battle. It was all around me. Suddenly, even though I was still furious, a sense of calm came over me. Guys who've been in battle often describe that strange feeling. It allowed me to analyze and assess the situation without emotion, keeping my anger inside. I started to look around, take note of every detail, and think of the best next decision. It was almost like going down a checklist.

Again, training was the key to surviving this fight. A split second after getting shot, and without even thinking about it, I identified the threat in the doorway, raised my rifle up to do a reflexive fire drill. But remember, my thumb was hit, so instead of putting my rifle on single fire, I inadvertently threw it on full auto. I have to say, that was the only guy that I've shot on full auto with my M-4. I normally shoot single fire because it's far more accurate, more controllable. But not this time; I dumped the rest of my mag into that guy. He was only about four or five meters away, max. It was a very close shot.

When my mag went dry, I then ran back behind the wall where I had prepped that hand grenade and did a quick mag change for my rifle. Now, normally, after going dry on my rifle I would have let it drop on the sling and instantly raised my pistol so I could have kept shooting, but for some reason I thought that taking time to change the magazine in my M-4, and having its power rather than a pistol, would be better during the fight. But that's not how we train. I was supposed to immediately transition to the pistol and keep the fire going, but in the moment, with the firefight being in the open and my guys still needing cover, I chose to stay on the rifle. It may seem minor to some, but that tactical detail, that decision, still bothers me to this day. I almost feel like I left Matt and Mike hanging. That wasn't the case, of course, but that decision—not to immediately switch to a pistol but change my rifle's mag—still doesn't sit well with me. The pistol could have kept the fire going without a pause, but the rifle had more firepower and accuracy. I guess it's a choice between two shitty options.

Regardless, at that point in the battle, I was thinking we should just Mogadishu those rooms—spray them with bullets and pull out whatever we find. It's not the way we do business, but I knew the bad guys were there, I knew where my guys were, and that there'd be no collateral damage, so I didn't want anybody else from our team getting shot.

So before I came back around the corner, a little voice in my head said, "Look to your right." And so of course I listened to the little voice, and on my right I saw all of the Iraqis who were supposed to be fighting with me. I'm like, "Oh, I was wondering where you guys wandered off to." They were frozen. At that point, probably because I was so jacked up on adrenaline and angry, I grabbed my Iraqis and started throwing them into the objective. Once I got some guys moving, the rest of them followed suit, and then they started spreading out and locking down doors the way they should.

Assault Force II linked up with our Iraqis and took control of the assault. The chief and Mattie, my other bravo, had taken control of all forces and finished securing the objective.

Our medic, Sergeant First Class Sean Howie, a good friend of mine, came over to take a look at us. He grabbed Mike and Matt and pulled them back behind a wall and started working on them. At that point, I thought to myself, "All right. My job's done."

I walked up to my chief and I said, "Hey, chief. I got shot."

He said, in a very matter-of-fact way, "I know."

We were talking as if it were a normal conversation, a normal evening out. That's what made the scene so surreal.

I said, "Okay, I'm going to go take a knee and pull security."

I walked to the outside edge of the perimeter, took a knee, and drank some water, faced out and started pulling security. Then I started cussing up a storm. [*Laughs*]

A naive question, maybe, but why weren't you lying down on the ground in pain after all of that?

Many other Special Forces guys have found themselves in similar situations, where they could have simply laid down … but you don't want to die. Sure, you're hurt. Yes, you've been shot, but you can't stop. If you stop, you die. If you've got time to pray, then you've got time to fight. Mike and Matt had been shot, too. They didn't stop. They were still fighting. They were still hitting dudes. I didn't do anything that they didn't. Matt said that getting shot felt like he had gotten speared playing football and couldn't move his legs. But he was still in the fight. He crawled, found cover behind the body of a dead terrorist, and shot a guy who could have killed me. The same thing with Mike.

He was lying in the middle of all this, seriously wounded, and couldn't even lift his rifle. Still, he didn't roll over and die, or even quit. He raised his pistol and kept shooting. Most in this profession would never surrender under similar circumstances. Anyone with the same amount of training and motivation, in fact, would probably do what we did, if not more.

What happened to the main target?

Obadiah got shot in the main room where those hand grenades detonated. That was unfortunate, because we don't go out to just kill people. We go out to capture, which is better because then we hand them over to trained interrogators. They can get information and we can keep working our way down the rabbit hole. I would much rather have valuable information than just a dead body.

What happened next?

I kept looking out from the perimeter. Damn, I can still remember what that field looked like. I was waiting to see if anybody else would come through, expecting to see a counter-assault force come from one of the nearby villages or houses after hearing the fight. If they were coming, I thought, then let them come. I was expecting it … and hoping for it, actually. More than anything else I was just pissed off, furious, angry ... not only from getting shot in the guts and shot in the thumb and blown up, but because Americans had gotten hurt. That wasn't how we usually did business. Usually our guys didn't get hurt. Usually we were good enough to take down the structure and catch the bad guys before they could even get a weapon into their hands. The fact that I saw two of my brothers get shot … at the end of the day, they did the right thing and they followed me. Throwing hand grenades in the room was like going towards the sound of gunfire. They had my back and I had theirs. Still, I couldn't help but feel responsible for them lying on the ground. I was angry, and that anger was probably the reason I was still moving around. Most of all,

I just didn't feel like I was finished. I didn't want the battle to end, and I didn't want to feel like I was tapping out, but then I had to balance all of that emotion with my professional assessment of the situation.

The medic came over and told me that the medevac bird was inbound. I probably said some colorful things to him. [*Laughs*] So he just let me be.

I looked back and forth to check on the guys to make sure it looked like they were doing the right thing, and they were. Like I said, Mattie and the chief took control. After that, it was just waiting for the helicopter.

Once the bird landed I got on, sat down, and threw my helmet in the corner. Once we were in the air and starting flying back I kept wondering, "Where are we going?" For some reason I wanted to go back to the team house. I wanted to go back to my room, which sounds kind of funny. Obviously, I was headed to the frigging hospital.

Once we were in the air for a little while, everything set in. In a moment it was like, "Hey, you're safe, and everything's okay now." The medic, Sean, had gone over me and was able to find the entrance wound but not the exit wound of my gut shot. He and another guy were holding gauze on my guts and trying to wrap my thumb. At first I was kind of fighting and cussing at them. Then shock set in, and I started getting sick. To this day I feel bad about it, but I threw up all over them. It was the veal parmesan that the 82nd Airborne cooks had prepared for us earlier in the night, and to this day I cannot stand the sight or smell of it, which is too bad because I used to really like it.

Meanwhile, poor Sean, he was working on Mike and Matt. They had Matt laid out across the seat so he could work on him, and then Mike crumpled up on the floor. He's working on both of those guys. I'm trying to throw up on the other side of the cabin so I don't get any on them, but I'm failing.

They must have radioed that it was a mass casualty event, because once we landed at the combat hospital there were probably twenty people at the airfield with stretchers and wheelchairs ... for three of us! [*Laughs*] Then I probably used my last five "cool guy" points. I got off the helicopter and I walked all the way to the last stretcher. I don't know why that was important in my mind at the time, but for some reason it was. It was like, "No. You guys worry about Matt and Mike. I'm going to the last one." I actually pushed some random male nurse aside because he tried to put me in a wheelchair. I said, "Get off me. I've got it." Sad to say, I wasn't a compliant patient. I got to the last stretcher, collapsed and I was done for the night. I even said it, actually. I yelled, "I'm done! I'm done!"

They wheeled me into the hospital. It was similar to what you see on television, with all the lights flashing above your head and people running around like crazy. The small world that it is, the first person who came up as they're wheeling me through the doorway was a very good buddy of mine by the name of Brian, who is an 18 Delta, a medic from one of our sister ODAs. He was at the hospital doing his trauma recertification. Now Brian's a surfer from California, a real calm, laid-back guy. So he walked up and said, "Oh! Hey, man. What's up?"

I said, "Brian! They shot me!"

Brian calmly said, "That's cool, bro." He started cutting my uniform off and trying to open the injury site while keeping me occupied. "Tell me more, Jar ... okay ... that's nice."

They put me under at that point and did the exploratory surgery. Apparently, because of all the pathogens or whatever that's in the air there, they would only do part of the surgery. If they had to do more, like pull your guts out, they'd send you to Germany to finish the surgery. That's what happened to us.

When I came to in Germany, I was on the operating table. I opened my eyes and saw all the tubes down my nose and throat. I picked up my head and leaned forward. I saw the doctors ... they were operating on me. They had their masks and gloves on, and there was this big white paper that they put around the incision site. I could see them reaching inside and moving things around. I even remember the sound.

One of the nurses looked over at me ... I still remember this so clearly ... she grabbed me by both shoulders and said, "Sir, you're still open. Don't move!" I nodded my head like it was the most normal thing, like, "Oh, yeah. That makes total sense." Then they hit me with more anesthesia and it was back to la-la land.

The next few days were actually pretty rough. Think about it; I had no idea what had happened after arriving at the hospital and two of my teammates were seriously wounded. I didn't know if they were alive or dead, and I had no way of communicating with the doctors or nurses because I had all sorts of tubes down my throat. My right thumb had been shot so I couldn't write, either. I kept coming in and out of consciousness, and every time I awoke I'd feel the concern for Mike and Matt, wondering where they were, or if they were okay, but all the nurses could do was tell me to calm down. Being unable to communicate was probably the most frustrating part of the whole experience. It really bothered me, emotionally. I finally learned that Matt and Mike were both okay, and we all eventually made full recoveries.

How do you feel about being awarded the Distinguished Service Cross?

Others who have received prestigious awards say the same thing, and it's usually something like, "I didn't do anything special. I was just doing my job." The reason people say that so often is because it's true. Medal of Honor recipients from the past like Frank Miller—one of my personal favorites from the Vietnam era—to paraphrase him, say, "Nobody wakes up this morning and says, 'Yeah! I'm going to go out and get the Medal of Honor today!'"

Life doesn't happen like that. Again, people like Frank Miller, who has been in some truly hairy situations, they don't think they did anything special. Even though when people read the stories like I did when I was a kid, they thought, "Oh my God! This guy is the baddest dude on the planet!" But to hear him talk about it, he doesn't think so. He thinks, "I got in a shitty situation. I got in a big firefight. That's it. I was lucky enough to survive."

Mike and Matt should have gotten the same exact award that I did (*Silver Stars were awarded to Captain Matthew Chaney and Sergeant 1st Class Michael Lindsay*). They did the exact same thing. We were all part of the same firefight. We all got shot up, fought a lot of guys, and the team ... *the team* ... killed the target. It wasn't due to one individual. One person doesn't go out and win a war. It's truly a team effort.

The Distinguished Service Cross is presented to Jarion Halbisen-Gibbs, May 2009 (*U.S. Army*)

Knowing what you know now, if you could go back in time to that
recruiter's office, would you do it again?

I'd do it again in a heartbeat. Not even a second question asked. I
would love to press rewind and do it all over again.

When guys fight together, they join a brotherhood. To this day,
I've got friends who are brothers in a way, who are closer than family.
I wouldn't trade their love and respect for the world. The only way
you get that is by going through the suck together. That's what it
comes down to. Yeah, you'll get a weird sense of humor out of it.
Some people might not understand you, or what you've been
through. To quote something I once read, "People sleep peaceably in
their beds at night only because rough men stand ready to do
violence on their behalf." If that's what it takes to keep my family safe
and to give them a good world to live in, that's absolutely a price I'll
pay. A price that a lot of other people will pay.

Jarion Halbisen-Gibbs was credited with single-handedly killing six
insurgents, including the high-value target. As Jar laid in a hospital in
Iraq, still reeling from the gunfight, the sun rose over Capitol Hill and
General Petraeus, his staff, Members of Congress, their aides, and the
media prepared for the landmark hearings. Many were shocked, however,
by the advertisement that awaited them in that morning's New York
Times. The far-left group Moveon.org had bought a full-page ad to attack
General Petraeus. Under his picture was the headline "General Petraeus or
General Betray Us?" with a sub-headline that read, "Cooking the books for
the White House." The ad's provocative headline was supported by a few
paragraphs of nonsense and outright lies. If the group had hoped to score a
few political points, or rattle the general, they failed. All sides condemned
the ad on the morning news programs and General Petraeus ignored it.
He and his fellow warfighters had spent the last four years dealing with
far worse than whatever punches Moveon.org could throw.

Shortly after 12 p.m., the hearing began, and America watched. "The
military objectives of the surge are ... being met," Petraeus said, after

telling the committee that his testimony wasn't shared with, or cleared by, anyone in the White House or the Pentagon. "In the face of tough enemies and the brutal summer heat of Iraq, Coalition and Iraqi Security Forces have achieved progress in the security arena. Though the improvements have been uneven across Iraq, the overall number of security incidents in Iraq has declined in 8 of the past 12 weeks, with the numbers of incidents in the last two weeks at the lowest levels seen since June 2006. One reason for the decline in incidents is that Coalition and Iraqi forces have dealt significant blows to al-Qaeda-Iraq. Though al-Qaeda and its affiliates in Iraq remain dangerous, we have taken away a number of their sanctuaries and gained the initiative in many areas."

After detailing some of the campaign's larger successes, Petraeus said that some of the surge troops could be withdrawn, but slowly and over many months so the hard-won gains weren't lost. Bottom line: the surge was a success, the extra troops could slowly, slowly, come home, but others would need to remain in Iraq to secure the peace.

"I believe Iraq's problems will require a long-term effort," Petraeus said. "A premature drawdown of our forces would likely have devastating consequences."

There would be no premature drawdown. Petraeus had won the day and the American public seemed convinced (in fact, two years later, a CNN poll found that 56 percent of Americans believed that the surge had been a success). Critics remained on Capitol Hill, however. During the next day's hearings in the upper chamber, then-Senator Barack Obama said that the impact of the surge was "relatively modest" and continued his opposition to that strategy, while then-Senator Hillary Clinton told the general and the ambassador that, "Despite what I view as your rather extraordinary efforts in your testimony both yesterday and today, I think that the reports that you provide to us really require the willing suspension of disbelief," and that "any fair reading of the advantages and disadvantages accruing post-surge, in my view, end up on the downside."

Every serious assessment of the situation in Iraq post-surge told a different story. Later that year in December, the Defense Department released their regular report to Congress titled "Measuring Stability and Security in Iraq," and it noted that "tactical and operational momentum has been achieved, and there have been notable overall improvements in the security situation. These improvements, combined with an increase in provincial government expenditure rates, have contributed to improvements in the delivery of essential services and other key programs to the Iraqi people. Cooperation with Iraqi and Coalition forces by tribal leaders—both Sunni and Shia—has advanced 'bottom-up' reconciliation and assisted in countering extremism."

As our warfighters like Staff Sergeant Halbisen-Gibbs kept fighting under their new strategy, the assessments became increasingly favorable. In late 2007 and early 2008, al-Qaeda wasn't only on the run in Iraq, they were running out of breath and collapsing from exhaustion. Their old hiding places were now occupied by American warfighters. Their old friends—or victims, rather—were sick and tired of their cruelty and dead-end ideas. The central government remained an obstacle, with its weak leaders and generations-old grudges between the factions, but our warfighters had done their jobs, and done them well.

"One year ago, extremists, cold-blooded killers, people who kill innocent men, women and children to achieve their ideological objectives were succeeding in their efforts to plunge Iraq into chaos," President Bush said in February 2008. "So I had a choice to make: Do I suffer the consequences of defeat by withdrawing our troops; or do I listen to my commanders, the considered judgment of military experts, and do what it takes to secure victory in Iraq? I chose the latter. Rather than retreating, we sent 30,000 new troops into Iraq, and the surge is succeeding. High-profile attacks are down. Civilian deaths are down. Sectarian killings are down. U.S. and Iraqi forces, who are becoming more capable by the day, have captured or killed thousands of extremists in Iraq, including hundreds of key al-Qaeda leaders—the very same people that would like to hurt America once again."

The new counterinsurgency strategy was working, but as its planners knew, the toll on American troops would be greater now that they were living among the people rather than holed up behind the walls of massive bases. On March 23, 2008, the number of military troops and Defense Department civilians who were killed in Iraq crossed the four-thousand threshold. A year earlier, such numbers would have intensified calls for a complete withdrawal, but because the American public was finally seeing results, and the war was being won, operations continued without interference from Washington. Then, as promised, by mid-July 2008 all of the extra brigades that were officially part of the approximately fifteen-month surge had departed Iraq, bringing our strength back down to about 147,000 troops.

Peter Mansoor had a front-row seat to the entire surge: its birth, its life, and its eventual retirement. The retired colonel puts the entire campaign into perspective in his book about the surge: "In January 2007 the ultimate outcome of the war was uncertain at best. Most onlookers gave the surge a small chance of success, if any at all. Yet against all odds, the surge succeeded ... By the end of the surge in July 2008, violence in Iraq had been reduced by more than 90%."

The surge of troops wasn't the only reason fighting was brought under control and al-Qaeda pushed to the margins. The foreign fighters who comprised many of its senior levels had worn out their welcome in the Sunni tribal areas. They had abused too many women, and their interpretation of Islam—cutting off the fingers of those who smoked cigarettes, for instance—was particularly unpopular in a culture that loves tobacco. Iraqis wanted them gone. Our troops were at the right place at the right time, and with the right strategy to build upon the growing anger against al-Qaeda among the population.

There was still much fighting to come, but our warfighters had turned the tide during the surge, and the long withdrawal of our forces from Iraq began. The only questions that remained were how long they would stay, in what size, and for what purpose. Getting those questions right would prove challenging, and decisions by a new president could either secure or squander the surge's victory. But for now, the effort was finally looking up.

PART IV

THE WITHDRAWAL, 2008-2011

CHAPTER FOURTEEN

"YOU COULDN'T HAVE PAID ME TO GO HOME."

NICK ESLINGER
SILVER STAR, *SAMARRA*

*I*n early September 2008, coalition forces officially gave Iraqi security forces complete responsibility for Anbar Province, which 2nd Lieutenant Diem Vo said was "inflamed with the insurgency" only two years before. *General Petraeus left Iraq a few weeks later to become the commander of U.S. Central Command, where he was charged with overseeing our military efforts across the wider Middle East, including the ongoing wars in Afghanistan as well as Iraq. His deputy, General Ray Odierno, was left in charge of our troops in Iraq. Robert Gates, the defense secretary who had replaced Donald Rumsfeld just before the surge began, travelled to Baghdad to preside over the change of command ceremony on September 16, 2008.*

"When General Petraeus took charge 19 months ago, darkness had descended on this land," Gates said after officially passing command from Petraeus to Odierno. "Merchants of chaos were gaining strength. Death was commonplace. Around the world, questions mounted whether a new strategy—or any strategy, for that matter—could make a real difference."

Gates went on to highlight how the new counterinsurgency strategy and the surge of troops had changed the course of the war. "Our enemies took a fearsome beating they will not soon forget ... the darkness has receded. Hope has returned."

Before departing, General Petraeus wrote an open letter to the troops who had made the surge a success.

"It has been the greatest of privileges to have been your commander," Petraeus wrote. *"When I took command of Multi-National Force Iraq in 2007 ... the situation in Iraq was hard but not hopeless. You have proven that assessment to be correct. Indeed, your great work, sacrifice, courage, and skill have helped to reverse a downward spiral toward civil war and to wrest the initiative from the enemies of a new Iraq."* General Petraeus then noted that the coalition's mission wasn't complete and that *"hard work and tough fights"* were still ahead.

One of our warfighters who'd be doing that hard work and continuing the fight was a young second lieutenant named Nick Eslinger. He had been in Iraq for about two months when Petraeus departed.

Eslinger was part of a new generation of American warfighters, one who grew up while our nation was at war and joined the military knowing full well that he'd be sent into combat. On the morning of September 11, 2001, Eslinger was still in high school. He and his best friend, Matt, were driving to morning football practice at Freedom High School in Oakley, California, where Nick was quarterback and Matt was a defensive end for the Falcons. *"It was still dark, about 6:10 in the morning, Pacific Time, when we turned on the car radio and heard what was going on,"* Nick told me. *"We got to the gym and the coach cancelled the workout. So we went back to Matt's house, had breakfast, and watched what was happening on television. The first time that I heard the name Osama bin Laden was that morning."*

A few months before the 9/11 attacks, Nick had decided to apply for admission into the United States Military Academy at West Point. A couple of cadets had visited his high school during his junior year and explained that West Point was *"a leadership factory."* That intrigued him. *"I had always been a leader on my sports teams, whether it was baseball, golf, football, or basketball,"* he told me. *"My peers always elected me to be a team captain. I enjoyed leadership and I felt really good at it."*

Eslinger learned he was accepted into West Point a few months after the attack, and wasn't at all apprehensive about joining the military during a time of war. So after graduating from West Point, the basic infantry officer school, and then Ranger School, the young platoon leader finally reported to the 101st Airborne Division at Fort Campbell, Kentucky, in late May of 2008.

Nick Eslinger with his mother shortly before graduating from West Point, May 2007 (*N. Eslinger*)

NICK ESLINGER: I was assigned as a platoon leader in Cougar Company, 2nd Battalion, 327th Infantry Regiment. The battalion's motto was "No Slack," and it was a very famous unit. The unit had already deployed. So I left in late June and, on July 4th, I flew from Kuwait into Iraq.

My platoon had forty-two men. As platoon leader, and the only officer in the platoon, I worked in coordination with my platoon sergeant, who was my right-hand man. My platoon sergeant was actually a staff sergeant, Christopher Anderson. He had not yet been promoted to sergeant first class, which is the typical rank of a platoon sergeant, so he was young for his position, but more than qualified. He had taken over as platoon sergeant about a month before I arrived. He was fairly new in the position, and I was very new in the position.

Our unit was based in Samarra, Iraq. We were at Patrol Base Olson, on the far west side of Samarra, along the Tigris River. A patrol base is typically a defensive position that you use for reconsolidation, reorganization, and mission prep. It was just a location that we used to sleep, eat, train, and prepare for missions. We had towers with weapon systems over-watching in all directions—360-degree security around the patrol base. I would say the patrol base was probably the size of two football fields in total.

The area was a hybrid urban/rural environment. The city had a couple hundred thousand people. We predominantly operated inside the city, which bordered the Tigris River, northwest of Baghdad, along Route Tampa, which was the main road in Iraq. North of Samarra you could see desert as far as you could look. To the east of the city was also desert, but south were more cities. To the west was lush farmland because of the Tigris River.

The city looked like nothing I had seen before in the United States, certainly. The monotone color of brown and dust was everywhere. It looked dilapidated throughout, from fighting since the war began. The roads had huge craters from roadside bombs. The homes were riddled with bullets. There wasn't a lot of infrastructure. Everyone was trying to steal electricity from one another, so the telephone poles and the wires were all just a mess. There wasn't a lot of cover from high structures, either. Everything seemed to be three stories or less, so not a lot of vantage points.

It was extremely hot, the hottest weather I had ever experienced. At one point on patrol, in July, I looked down at my watch thermometer and it said, "138." It sounds like your blood would just start boiling. You couldn't be on the ground more than thirty minutes in your gear before we had to rotate people into the vehicles for air conditioning. You've got your body armor, your vest and plates, which weigh thirty to thirty-eight pounds. You've got your Kevlar helmet, which is about three pounds. You've got ammunition, which probably adds ten pounds. So you're looking at anywhere between forty and fifty extra pounds, at least, and that's for a rifleman. For a soldier who is a grenadier, the grenades weigh more than bullets so he's carrying more weight. And the medics are carrying aide bags on top of whatever else they're wearing.

My first impression of the city was that it was safer than expected, because back at the rear detachment I had heard that a lot of soldiers were still dying. There was still a great deal of fighting going on. When I got there, that didn't seem to be the case. When I talked to other platoon leaders, they said, "Oh, six months ago there used to be attacks every day in Samarra. But recently, it's maybe one a week or one a month." It was very calm; at least that was my first impression. I thought, "Maybe I won't see as much fighting as I thought ... at least initially."

Most of the gains in security occurred in the spring and summer of 2008, and by the end of the summer our company was being attacked on average 1–3 times a month, which was drastically reduced from the 75–100 attacks per month at the beginning of 2008. I credit the stable security situation to two factors. First, and most appropriately, I credit our unit's effort and grit to remain constantly vigilant and present throughout the streets of Samarra. I have no doubt that our twenty-four-hour presence was the greatest deterrent. The second factor was the Iraqi security forces. The joint effort of the Sons of Iraq [*a friendly militia that grew from the Anbar Awakening*], the Iraqi National Police, and the Iraqi Army finally reached a level of effectiveness that the enemy had to respect. Their increased legitimacy in the city of Samarra assisted our U.S. efforts to maintain a good security situation.

Nick Eslinger stops his patrol in a market in Samarra, September 2008. (*U.S. Army*)

How old were you?

I was twenty-four, and absolutely loved my job. I was doing what I was trained to do and I felt comfortable doing it because West Point had prepared me. I didn't want to leave or be anywhere else other than on patrol with my men. You couldn't have paid me to go home. I wanted to be with my platoon in Iraq.

Why?

Try to think of it in sports terms. For years ... literally for years ... all you've done is practice, practice, practice. Finally, when you get to deploy and do what you practiced for, you're finally in the game. That's all you've wanted ... to get off the bench, stop practicing, and get in the game. Of course, then you don't want to leave because you know you just have to go right back to practicing. You have to go back to training and rifle ranges, and dealing with all the things in garrison. You don't want to do that. You've done that for years. Through West Point, through Ranger School, everything has been preparing you for this moment to deploy and lead, or "get in the game." So now you're there ... and you don't want to leave.

Tell me about that night in October 2008.

At that time we had had a threat in the city called an RKG-3. It's a cylinder-shaped device with a handle at the bottom of the cylinder, maybe a foot and a half long in total. It's got a little parachute on the end of the handle to stabilize the device so that when it impacts the target, the end of the cylinder is the point of impact. Then it shoots a slug, penetrates the armor, and bounces around inside of the vehicle to cause a lot of damage. It has a lot of heat. We like to say it "travels at the speed of light with the heat of the sun." Those things are dangerous, and they're scary. They killed people during our deployment, as well as destroyed vehicles down to nothing by setting them on fire and all we could do was watch them burn because of the risk posed by the unspent ammunition still inside the vehicle.

Anyway, we had a few RKG-3 attacks in Samarra. So when we went on patrol, we were knocking on doors, talking to people and trying to gather intel about where these things were coming from. We'd ask, "Who's using them? Do you know anyone that has one right now? How are they using them? What are their techniques? How are they deciding what to throw at?" Just general reconnaissance, trying to paint the picture so we could come up with a plan to defeat them.

So at that time, most of our patrols were focused on reconnaissance regarding RKG-3 grenades. My commander, Joshua Kurtzman, put the platoons on a rotating schedule. We were either in the city patrolling, guarding the base, or out on what we called "long-distance patrol," which was on the periphery of the city. Whichever platoon was on long-distance patrol cycle didn't come in the city, but patrolled the farmland and deserts outside of the city. We rotated every four days. On October 1st, my platoon was on the Samarra rotation, so we were inside the city.

When we were on city rotation, we had a day patrol and a night patrol. The commander expected us to be out on each patrol any-where from four to six hours during the day, and two to four hours at night. It was up to the platoon leader to build the task and purpose for each patrol mission. I could literally go wherever I wanted in the city. We decided that morning that we were going to stay north. Then in the evening, we were going to go southeast of the city to a neighborhood called Jiberia II. When we got back from the morning patrol, nothing significant had happened. The platoon sergeant and I went back to our CHU [*containerized housing unit*] to plan the next mission and the platoon went to go recover, work out, sleep, or whatever they needed to get ready for the night patrol.

I talked to my platoon sergeant and we came up with the plan for the evening patrol. We were going to do more reconnaissance, knock on some doors, and talk to some people that we hadn't yet talked to, and just try and gather more intel. At night, it was a little bit easier to knock on doors because most people were home, versus in the day when they were either at work or traveling to see friends. Not a lot of people answer the door during the day. We knew at nighttime we'd get some answers; we'd see some people.

We ate before we left. I'd say we SP'd, which means the "Start Point" of our patrol, about 1900 that night in three MRAPs [*Mine-Resistant Ambush Protected Vehicles*]. It was sunset when we rolled out, and it gradually got dark. We stopped at the Iraqi police base and picked up a squad of Iraqi policemen to assist in reconnaissance. In total, travel time was probably twenty-five minutes.

We entered Jiberia II and got to our preplanned dismount point on the periphery of the neighborhood. I was the dismounted patrol leader, and my platoon sergeant stayed with the mounted patrol element. If something were to happen, I wanted to be at the point of friction, which was more than likely to occur on the ground.

I dismounted with one of my squads and six Iraqi policemen. Staff Sergeant Samuel Heath was the squad leader, and in total we had around eighteen guys on the ground with us. Another squad with my platoon sergeant manned the vehicles and screened around our position. They continued to circle and warn us about things they saw. They were also close enough to where if something happened they could quickly come reinforce us with the vehicles.

The dismounted patrol started on our route. We knocked on a few doors, sat down, and talked. We didn't really get any intel at that point and continued moving along. People on the street were apprehensive about coming near us. That made sense because of our appearance, and they knew that we probably weren't there for anything good. So they saw us coming and they'd go inside their house. The people who actually would let us in their house would be very guarded. I think they were afraid that there would be consequences ... significant consequences ... if they told us anything useful.

They would always put the women and children in a separate room and we would meet the elder of the household, or the man of the house. The wife would always bring us tea. They would be very hospitable, but out of necessity. They felt they had to do it, not that they wanted to do it. Generally, we would receive one-word responses or very vague made-up stories. It was easy to tell that they were just trying to give us lip service and get us out of their house. I wouldn't say we were received well.

It was one of those nights when I told my platoon, "You can decide whether you want your NODs [*night optical device*] up or down," because in Jiberia II there was enough ambient light from the moon and street lamps that you could see better without NODs. Later that evening, having my NODs up saved my life.

About 19:45, we found ourselves in a sort of alleyway, a long dirt street about ten yards wide. On each side were a series of houses with courtyards in front, and walls, about seven or eight feet high, separating the courtyards from the street—that's why it looked like an alley.

I don't remember people being in the street. There were some cars parked along the side of the road, but mostly it was open. I was on the left-hand side with my side of the patrol. The other guys were on the right-hand side—about nine on each side, spread out. I was in the middle, with my radioman and medic right next to me. We were moving along and I pointed to my right, towards the gate of the next house where I wanted to stop. We came to a stop and we each took a knee and looked around. Then the squad leader moved over to the courtyard gate, knocked on it with the interpreter, and then requested permission to come in and talk to the family.

I had taken up a position next to a wall alongside the street. To my front, there was an open lot where a house should have been. It was roughly twenty to thirty yards wide, and about twenty to thirty yards deep. It was just this big square open lot. There was a wall on three sides, except the side that faced the street. They had neighbors to each side, but no house there, just an open dirt lot.

I was on a knee on a corner of that lot, and because of the terrain it was my responsibility to cover that sector. So, I was pulling security but I also had a responsibility to maintain communication with my guys. I had my NODs up. I looked down, because my radio microphone was on my left shoulder, and said into the mic, "Hey, do we have permission to enter the house?"

When I let go of my microphone and looked back into my sector, I saw in the corner of the lot, about twenty to thirty yards away, a hand come up over the wall and throw a projectile. It looked like a rock. It left his hand and came towards our patrol in the street. I instantly thought "grenade" because of the shadow or the shape.

I yelled, "Somebody threw something!"

I don't know why I yelled that ... why that was the sentence that came out of my mouth ... but it was. As I yelled that phrase, I kind of stood up from kneeling, not all the way, but halfway, and pushed off on my left leg and started going to my right. I dove to the point where I thought I could intercept or catch the grenade, to where I hoped it would land. I then found myself on the ground simultaneously landing on my right side when the grenade hit the ground, and it rolled right into, and then underneath, my side, hitting my rib cage area against my vest.

Over the years, a lot of people have asked me why I did that. Honestly, it felt like a reflex. It felt the same way it does now when I have to stop suddenly in my car and I reach out my right arm to protect my wife in the passenger seat. Just a reflex. I mean, there's a grenade falling through the air toward your men ... what else are you supposed to do? Take ten seconds to weigh your options and do a risk assessment? No. Of course not. You do not have time to ponder, "Do I run? Do I yell? What do I do?" You simply do not have time. You revert back to instinct to protect your guys because that is what is expected of you. That is what a platoon leader does—accomplish the mission and protect your men. I guess that is what my brain was thinking at the time. I just dove, for no other reason than that.

Nobody else saw the grenade come over the wall because they were all focused on their individual sectors. Later on, my medic told me that he thought I was wrestling with somebody on the ground. He was like, "Why is the lieutenant on the ground?"

I felt the light thump of the grenade under my ribcage, when it hit my vest. I had both hands free at that point because I had let go of my weapon when I dove. I started reaching for the grenade with my right hand, and by the grace of God, I put my hand on it and picked it up. It was a palm-sized grenade. I think it was one of those pineapple-like grenades used in previous wars. I do remember that it wasn't smooth. Although I only gripped it for a split second, I'll never forget the way it felt in my hand.

Anyway, I somehow managed to hurl the grenade back toward the wall it came from.

You played shortstop when you were a kid, right?

Yeah, my baseball coach told me once that I had good dexterity. [*Dexterity is defined by Merriam-Webster as "the ability to use your hands skillfully, the ability to easily move in a way that is graceful, and the ability to think and act quickly and cleverly."*]

I tried to yell "grenade!" as I threw it, but the noise and concussion when it detonated took my breath away. I don't know if it detonated in midair or when it hit the ground again, but none of us received any wounds.

My guys were all surprised. They didn't know what happened because I was the only one looking in that direction. It happened so fast, they just heard my voice and the blast and were wondering what the heck just happened.

That was my very first "contact," as we call it in the military. My heart was trying to burst through my body armor, but on the outside I was just trying to give directions as calmly and collectedly as I could. I directed Staff Sergeant Heath to take half his squad and go back down south, wrap around to the next street over, and try and see if they could catch the guy that threw it. Meanwhile, the other half of the squad and the Iraqi policemen secured our current location. Before I told him to, my radioman had already called the platoon sergeant, told him what happened, and said to bring the vehicles toward us. We were all checking each other to make sure we were not wounded. I sent doc [*the medic*] around to each person, making sure everyone was fine.

How much time passed from the moment you saw the grenade being thrown to when you threw it back?

Three seconds ... maybe. It just landed right where it needed to. If it had landed anywhere else, anywhere else at all, the result would have been very different. I hate to use the word "perfect," because it's a situation where you shouldn't use that word, but it was a perfect landing and I was in a perfect position to grab it and throw it back before it detonated.

If the grenade had continued rolling, I think it would have rolled to the far side of the alley, where half of my patrol was standing. It would've been in very close proximity to about five guys: my squad leader, the Iraqi policemen squad leader, my interpreter, my SAW gunner, and a rifleman. About five guys would have been within eight yards of it. That's something I'm not happy admitting because you're supposed to maintain spacing between your men. But you've got to remember, at the time, we were converging on a courtyard gate because we were just about to enter that courtyard. Guys were getting up and moving towards the gate, which is a funnel point. That was a perfect target, and that's probably why he threw it at the time he did. A lot of soldiers were in one spot, and he could inflict the most damage. An American fragmentation grenade has a kill radius of five meters and a wounding radius of fifteen meters. I don't know whether this grenade was more, less, or equally powerful, but if you're within about ten meters of any grenade detonating you're more than likely going to be killed or seriously injured.

So, we next went to the location of the throw, which came from the backyard of a house. On the outside of the house was a drawing, in chalk. It was like those Native American cave paintings of things that happened in battle. It was just like that, but of an American patrol, and it was an illustration of an attack on Americans. It had a little drawing of flames where the explosion would happen. It had American vehicles, too. It was bizarre. We also found the spoon and safety pin from the grenade in the dirt. We sent them off for analysis, but unfortunately they had no identifying prints or marks.

While still at the house with the backyard the grenade came from, a woman approached us and said, "My son was the one who threw it." She also told us where he worked. That's how we found him the next morning. We went to where he was working, talked to the boss and said, "Where is this guy?" He pointed to him; we sent a squad over there and detained him.

I do not regret what I did that night. Quite the opposite, actually.

You know, some may think it was traumatic enough to have lasting negative effects. Well, it doesn't affect me, or haunt my dreams, or anything like that. It certainly was feasible, and probably would have been understood by many, including my commander and my soldiers, if I had just yelled "grenade" as loud as I could and tried to take cover. But that's just not how it turned out.

Nick Eslinger inside his containerized housing unit at Patrol Base Olsen in Samarra, 2008. (*N. Eslinger*)

Second Lieutenant Nick Eslinger received the Silver Star for his actions that evening. The citation reads that he "took actions under extreme circumstances that saved the lives of his soldiers" and that "his initial instinct to sacrifice himself for his comrades is truly valorous."

I did what my commander and my soldiers would have expected me to do. I think that if anyone else on the ground that night had seen what I saw, been in the position I was in, in terms of proximity to affect things, they would have done something similar. Had I dove the other direction to get away from the grenade, or done something different and allowed the grenade to land and roll toward my squad, I don't think I would have been able to live with myself. Families that exist today may not have ever existed if I didn't do what I did. Then again, my marriage and two daughters wouldn't exist if I had died on that grenade. Who knows if that's true, but when I see pictures or hear about the families of my soldiers on the ground that night, I feel an overwhelming sense of pride. It's a sense of pride a thousand times stronger than I feel from a medal on my chest.

Eslinger and his fellow warfighters remained in Iraq fighting insurgents, protecting the population, and securing neighborhoods so that vital public services and institutions could return. The positive trends continued into the next year, and as things progressed, so did the plans for coalition warfighters to withdraw. President Obama had promised many times on the campaign trail to leave Iraq, once calling it a "dumb war." He was inaugurated on January 20, 2009, and less than forty days later he announced the results of his national security team's review of our strategy in Iraq.

"The United States will pursue a new strategy to end the war in Iraq through a transition to full Iraqi responsibility," Obama said during a speech at Camp Lejeune, North Carolina. "Let me say this as plainly as I can: by August 31, 2010, our combat mission in Iraq will end ... I intend to remove all U.S. troops from Iraq by the end of 2011."

A timetable for withdrawal had been set in stone and many thought it was a reckless decision. Critics said a firm date would empower our enemies to simply wait us out. They could lie low until we left and prepare to attack once we were gone. We might depart, but the war would continue. "No war is over until the enemy says it's over," U.S. Marine Corps General Jim Mattis said earlier that year. "We may think it over, we may declare it over, but in fact, the enemy gets a vote." Others saw the artificial deadline as somewhat inevitable; "We can't stay in Iraq forever," they'd argue.

In late June 2009, the Iraqi state television began running a "Countdown to Sovereignty" clock, showing the dwindling hours until the moment when all U.S. forces were scheduled to close the last of their neighborhood-based combat outposts and return to their bases. An article in the June 30, 2009, edition of the New York Times captured the moment: "Iraq celebrated the withdrawal of American troops from its cities with parades, fireworks and a national holiday on Tuesday as the prime minister trumpeted the country's sovereignty from American occupation to a wary public." The mood was celebratory, but some still had doubts. "Some people are afraid because the Americans have left," one female Shia said, adding that "some think it will be better because then the enemies of the Americans will leave Iraq" and the attacks aimed at derailing their strategy would stop.

A few months later, on April 19, 2010, the two most wanted terrorists in Iraq were killed, al-Qaeda in Iraq leaders Abu Ayyub al-Masri, who took over the group after al-Zarqawi was killed in 2006, and one of his top deputies, Abu Umar al-Baghdadi. The commander of our forces in Iraq, General Ray Odierno, issued a news release saying that the "death of these terrorists is potentially the most significant blow to al-Qaeda in Iraq since the beginning of the insurgency."

Less than a month later the Associated Press reported that a group known as the Islamic State in Iraq, which was described as an umbrella group that oversaw several insurgent networks, including al-Qaeda in Iraq, had named a replacement for al-Masri. His name was Abu Bakr al-Baghdadi. Few paid any attention to this new leader of a beaten and presumably broken network, but al-Baghdadi would eventually come to rule over one of the region's most brutal "states" in centuries: the Islamic State of Iraq and Syria, or ISIS.

CHAPTER FIFTEEN

A TIME OF PEACE?

President Obama ordered our last "combat troops" to depart Iraq in the summer of 2010, leaving a "training force" of fifty thousand that would continue to dwindle as the months passed. It was a vague distinction between terms, though. As most warfighters would say, a soldier is a soldier, and there's small difference between combat troops and a training force when you're still very much in a war. Still, the distinction allowed the president to keep his campaign promise, and more importantly, pave the way for the United States to completely withdraw the following year.

"Will all (our) efforts pay off?" asked Lieutenant Colonel Darren Wright of Dallas, Texas, who was second in command of the last combat unit to depart Iraq, the 4th Stryker Brigade, 2nd Infantry Division. Wright spoke to a *Washington Post* reporter on August 19, 2010, before his unit convoyed south across the Kuwaiti border. "Of all the time and effort that we put in this country, the blood, the sweat, the tears, I wish you could see an answer within a couple of weeks or a couple of months."

Some believed they already knew the answer to Wright's question, however, including an enlisted man in his brigade, Specialist Clinton J. Clemens of Edgefield, South Carolina. Then twenty-six years old, Clemens was only eighteen when he was part of the invasion force in 2003. Now, as he and his fellow warfighters left after seven years of hard fighting, he was less optimistic about the war-torn nation

America was leaving behind. "I hope good things come from it," Clemens told the *Washington Post* reporter. "But I think as soon as we leave, things are going to fall apart."

The Defense Department's quarterly report issued that same month showed continued hope as Iraq improved across all measurable sectors— political, rule of law, economic and energy, and security trends. Despite this assessment, the study also said that "although stability is improving, it is not yet enduring" and the paper was brimming with language that painted the Iraqi security forces as unable and unprepared to fight against insurgent forces without at least some coalition support. These observations were at odds with the White House's strategy, and many at the Pentagon and elsewhere hoped the president would see this and eventually adjust his firm date of departure.

Meanwhile, Jarion Halbisen-Gibbs, then a sergeant first class, finally had the bullet extracted from his hip in 2010 that he had received three years earlier in that pre-dawn gunfight. (Some friends had the bullet engraved with the Special Forces crest, the numerical designation of the Operational Detachment Alpha team—083—and the date it ripped into his body: "10 SEP 07.") Doctors had worried that removing the bullet would cause more damage, but the Green Beret worked hard through rehabilitation and was eventually cleared to return to duty. He kept in contact with his Special Forces team while he was in recovery; they had returned to Samarra. "I couldn't believe some of the things I was hearing. Samarra was quiet. No IEDs. No harassment fire. The religious sites were still intact."

Halbisen-Gibbs said he was determined to get back to his team. He returned to Iraq in November 2010 to see the "dynamic change" for himself. The combat veteran was amazed. "Samarra had drastical-ly changed since the last time I was there," he said, speaking of the relative peace of late 2010 compared to the spring of 2007, when the al-Askari Shrine was bombed for the second time and months of heavy fighting ensued. "The city and the surrounding areas used to be the Wild West. Route planning was crucial to survival, and weaving our way through the city to avoid IEDs and ambushes was just par for the course. Enemy contact was never a question of if, but when," he said. "But during my deployment from 2010–2011, I saw a very different Iraq. I saw a nation that was rebuilding and changed for the better. A nation that was moving forward, not back."

A hard-earned souvenir: the bullet removed from Jarion Halbisen-Gibbs
(*J. Halbisen-Gibbs*)

Signs of hope sprouted all across Iraq. Bombings did continue, and at a rate that would have dramatically shook most nations into a state of panic. But in a country that had felt the brunt of war since the early eighties, and in a culture that had always lived in a relative state of conflict, a few bombings weren't really news. In fact, the West didn't consider it news either. Iraq had all but fallen from the front pages of our newspapers and the leading segments of our television news programs. The question of "What's next in Iraq?" took a backseat to domestic issues like the economy and the president's massive healthcare regulatory proposal, and a renewed, yet still low, interest in Afghanistan, which has since become the longest war in American history.

Even though President Obama had promised that American troops would completely withdraw from Iraq by the end of 2011, most Pentagon officials were still hoping that he'd change his mind in the final months of the year. Many believed that a few thousand troops could provide the intelligence collection, advice, and contin-ued training that would help retain the hard-won gains in recent years. A few thousand troops wouldn't necessarily commit the United States to endless fighting, but they would give us a place at the table where we could guide Iraq's fledging leadership in the coming years. If we completely left, they feared, there'd be no reason for anyone in Iraq to listen to American advice, sectarian squabbles would emerge, and all would eventually fall apart.

Leon Panetta, a lifelong Democrat politician who was chosen by President Obama to become his Secretary of Defense after Bob Gates left, was one of those who asked the president to leave a small force in Iraq. But time was running out. "Now that the deadline was upon us, however, it was clear to me—and many others—that withdrawing all our forces would endanger the fragile stability then barely holding Iraq together," Panetta wrote in *Time* magazine years later.

There was also a major sticking point, something called a "Status of Forces Agreement," or SOFA, which governs the legal status of our warfighters while serving in a foreign nation. Mainly, we needed our SOFA to give American troops immunity from local prosecution. It stated that our military would handle all of our own prosecutions and punishments for illegal actions. Without it, our troops could be hauled before kangaroo courts in every jurisdiction of Iraq, where show trials could condemn them to unimaginable punishments. As a matter of long-standing policy, America doesn't send its troops into such a questionable legal environment. Our military had immunity since the invasion, but it was due to run out at the end of the year because the president hadn't negotiated any continued presence.

Without a SOFA granting immunity to our soldiers, any discussion of a troop extension was moot. Rather than immunity being granted by the executive powers of Iraq's prime minister, Nouri al-Maliki, the question was sent to the nation's parliament, where it was sure to fail. Negotiations collapsed.

"We had leverage," Panetta wrote, amid a growing—yet uselessly retrospective—debate about the troop withdrawal. "We could … have threatened to withdraw reconstruction aid to Iraq if al-Maliki would not support some sort of continued U.S. military presence. My fear …

was that if the country split apart or slid back into the violence that we'd seen in the years immediately following the U.S. invasion, it could become a new haven for terrorists to plot attacks against the U.S."

The president ignored the advice coming from Panetta and his top commanders and refused to press al-Maliki for a deal. By mid-October, all hopes for a residual troop presence in Iraq had officially vanished. "A senior Obama administration official in Washington confirmed Saturday that all American troops will leave Iraq," read an *Associated Press* report filed in mid-October 2011. "The decision ends months of hand-wringing by U.S. officials over whether to stick to a Dec. 31 withdrawal deadline that was set in 2008, or negotiate a new security agreement to ensure that gains made and more than 4,400 American military lives lost since March 2003 do not go to waste."

The next month, on November 14, 2011, a roadside bomb in Baghdad killed a twenty-three-year-old paratrooper in the 82nd Airborne Division from North Carolina named David Emanuel Hickman. "He was supposed to come home December 1, but he didn't tell me because he wanted to surprise me," said his mother, Veronica Hickman, in an interview with the *Los Angeles Times*. "And I didn't tell him I had a surprise party planned for him. And now ... my surprise turns out to be that I'm going to my son's funeral."

Hickman, an all-conference linebacker for the Northeast Guilford High School football team, was the 4,474th American warfighter to die in the Iraq War ... and the last. As his body was laid to rest in Lakeview Memorial Park Cemetery in Greenville, North Carolina, his fellow soldiers were pulling out of their bases across Iraq and driving in their final convoys south across the Kuwaiti border.

"Today, I've come to speak to you about the end of the war in Iraq," said President Obama on December 14, 2011, during a speech at Fort Bragg, North Carolina. "Tomorrow, the colors of United States Forces-Iraq—the colors you fought under—will be formally cased in a ceremony in Baghdad. Then they'll begin their journey across an ocean, back home."

Four days later, under the cover of night, the last convoy of American troops left a small base near Nasiriyah, the city that Justin LeHew and his

Marines had fought so hard to take nearly nine years earlier. The vehicles crossed the border around dawn, and when the sun rose over Iraq, the only remaining American troops in Iraq were a few dozen guarding our embassy in Baghdad. Iraq's security forces, with all of their flaws, now stood alone to fight what remained of the insurgency. Many Iraqis cheered, celebrating as passionately as they had when our troops pushed Saddam's forces into retreat years before. Others worried. "Things will go worse in Iraq after the U.S. withdrawal, on all levels—security, economics and services," said Hatem Imam, in a *New York Times* article about the last American convoy to leave his nation. "We are not ready for this."

America's warfighters had won the war, but it was now up to the Iraqi people to win the peace.

Back home, Eric Geressy was one of the relatively few Americans who were watching news coverage of the historic withdrawal. Even today, many could tell you where they were when they watched the first hours of the 2003 invasion on television, and some might even be able to tell you the date, but most have no idea of when we actually left Iraq. "It just ended like it had never happened, without anyone paying attention to it," said Geressy, by then a sergeant major. "It just did not seem like it ended right."

The television coverage showed the final vehicles leaving, followed by a few comments from journalists ... and then it was over. Geressy and others didn't know what they expected to see, but it was certainly something more worthy of the sacrifices made during the war. Americans may have supported their warfighters, but they were more than ready to forget the war itself. Its nearly silent conclusion gave credence to the fundamental critique of how we chose to fight the conflict: our nation didn't go to war—only our military did.

"The military members and their families were the only ones who shouldered the burden of that war, and they did it almost completely by themselves," Geressy said. "The troops endured multiple tours of duty, year after year, while their families sat home waiting and hoping to never get that knock on the door. And now, the only people who remember, and understand, are those who served alongside their brothers and sisters ... on their left, and on their right."

More than 1.5 million American warfighters served in the Iraq War from March 2003 until December 2011. In nearly nine years of fighting, 4,490 Americans, both warfighters and civilian personnel, died there, and nearly 32,000 were wounded. Our warfighters committed countless acts of valor on the battlefield, where courage, and love for their brothers and sisters in arms, broke through the fires and fogs of war and allowed the very best of America's heart and spirit to shine in the desert sun.

Their service, their sacrifice, and the valor they displayed in Iraq destroyed a great evil, without a doubt. Their tireless efforts—literally their blood, sweat, and tears—also gave the people of that war-torn nation a chance to live in a democracy, at peace with itself and its neighbors. But it was now up to the Iraqis to determine the future of their home, and the world watched with hope that the coming years would be, as the book of Ecclesiastes promises, "A time of peace" in Iraq.

But it just wasn't so.

EPILOGUE

"Only the dead have seen the end of war." — Plato.

On the Sunday morning in August 2015 when I finished the first draft of this book, the above-the-fold headline in my local newspaper read, "'Perverted jihadist' won't shake military's resolve, Biden says."

The vice president had been speaking at a memorial service in Tennessee for five of our warfighters who had been recently murdered by a Kuwaiti-born Islamic extremist who attacked their reserve center in Chattanooga. During the ceremony, one of their officers spoke of the valor shown by the fallen during the attack: "When many of us would run, their gut reaction was to protect and ensure the safety of their brothers and sisters."

It was a typical response from our warfighters who have been fighting Islamic terrorists like that for nearly fourteen years in Iraq and Afghanistan. Now, they were fighting them in America.

News like this is becoming all too frequent. It seems every week brings reports of an American citizen joining ISIS, and every month we see news coverage of a planned or even a successful attack somewhere in the West, and now even in the United States. The incidents on our soil remind me of what President Bush told Aubrey

311

McDade ten years ago. "This fight has to be fought," the president said. "It might be fought here or it will be fought there, but it's going to be fought."

After our troops departed Iraq in December 2011, ISIS launched a blitzkrieg from their strongholds in war-ravaged Syria and into Sunni-dominated Iraq, filling the security vacuum created after we left. ISIS quickly took control of Mosul, where Eric Geressy and his 101st Airborne Division had established peace early in the war. Then they took Fallujah, where Aubrey McDade and his Marines fought one of the war's bloodiest battles. They took Ramadi and Hit, where Diem Tan Vo helped launch the Anbar Awakening. They're launching attacks into Baqubah, where Chris Waiters and Jeremiah Church decimated the enemy during the surge, and ISIS is bombing sites in Samarra, where Jarion Halbisen-Gibbs and Nick Eslinger saw hard-won measures of success after years of fighting.

The invasion was a success, but it was followed by years of an insurgency that wasn't defeated until we adopted a winning strategy. The war ended as a relative victory, until our premature departure paved the way for ISIS.

In Iraq, we lost our way, we found our way, but then we walked away. ISIS stormed in behind us, and they're worse than anything we've seen in decades, centuries perhaps. The sad thing is that many weren't surprised by their rise to power.

"This was something you could see coming," said now-retired General David Petraeus on June 30, 2014, at the Aspen Ideas Festival. "This grew out of al-Qaeda in Iraq. Now, at the end of 2011 when our final combat forces left, al-Qaeda in Iraq was … defeated. Over time they were able to resurrect a bit. The pressure was not kept on them in the way that it might have been."

General Ray Odierno, who was once deputy to Petraeus and then the overall commander in Iraq, agreed. "It's frustrating to watch it," Odierno said in an interview shortly before retiring in the summer of 2015. "I go back to the work we did in 2007, 2008, 2009, and 2010 and we got it to a place that was really good. Violence was low, the economy was growing, politics looked like it was heading in the right direction." The general's usually stoic expression during the television interview cracked slightly, and viewers could tell they were watching a man who was genuinely sad about the outcome. "If we had stayed a little more engaged, I think maybe it might have been prevented," Odierno continued. "I've always believed the United States played the role of honest broker between all the groups and when we pulled ourselves out, we lost that role."

In many ways, Iran also filled that vacuum, fueling the Shia-Sunni divide and influencing the prime minister to further push his rivals from power. Experienced Sunni generals were purged from the military, replaced by untrained and unprepared Shia loyalists. Ministers and government employees met the same fate. It was like the disestablishment of the army, de-Baathification, and sectarian violence all over again, and the Sunni response was much the same. The shunned Sunnis turned to the extreme elements within their communities for protection from the government and the Shia death squads, but instead of finding former Baath hardliners, they found the twisted, evil Islamic extremists who were left in the wake of the insurgency's defeat.

Is ISIS worse than Saddam's Baath party and al-Qaeda? Yes. All those who oppose ISIS are executed, often in grotesque displays of public torture and with medieval methods of killing. Members of the Iraqi security forces or local militias who are captured—or often simply men and boys who are of fighting age—are decapitated, shot, or blown up. Children sometimes perform the executions. Journalists and aid workers are beheaded; captured fighters are burned alive and drowned in cages before cheering crowds. Homosexuals are thrown from the roofs of tall buildings. Sharia law has been established, ancient artifacts deemed un-Islamic are being destroyed, and women and girls have fallen into a system of slavery and systemic rape.

Last summer I read a disturbing article in the *New York Times* titled "ISIS Enshrines a Theology of Rape," by Rukmini Callimachi that exposed the horror.

"In the moments before he raped the 12-year-old girl, the Islamic State fighter took the time to explain that what he was about to do was not a sin," Callimachi wrote, describing an interview with a young girl who had escaped from ISIS. "Because the preteen girl practiced a religion other than Islam, the Quran not only gave him the right to rape her—it condoned and encouraged it, he insisted." Callimachi went on to describe how ISIS had created not only a religious justification for this abuse, but a complete bureaucracy to handle the backlog of slaves: courts, clerks, guards, and warehouses. Guidelines were established and a how-to manual was even produced, citing the Quran and other teachings that not only allow this to happen, but require it to happen in order for the Islamic fighters to become more virtuous.

"The youngest, prettiest women and girls were bought in the first weeks after their capture. Others—especially older, married women—described how they were transported from location to location, spending months in the equivalent of human holding pens, until a prospective buyer bid on them," Callimachi wrote. "Their captors appeared to have a system in place, replete with its own methodology of inventorying the women, as well as their own lexicon. Women and girls were referred to as 'Sabaya,' [slave] followed by their name. Some were bought by wholesalers, who photographed and gave them numbers, to advertise them to potential buyers."

ISIS claims some sort of religious mandate for their actions, but it's clear to everyone that their conquest is nothing but the brutal rape and plunder of centuries past. It's a great shame upon the entire civilized world that they're allowed to exist.

Two days after Callimachi's story was published, the *Associated Press* reported that an American hostage—twenty-six-year-old Kayla Mueller of Prescott, Arizona—had been sold to the leader of ISIS, Abu Bakr al-Baghdadi. The aid worker had come to help the Syrian people; instead, she was enslaved, sold, and her fingernails were ripped out before she was repeatedly raped and tortured. She eventually died during an airstrike on ISIS fighters in early 2015.

"The report (about Mueller) was extraordinary on so many levels it's almost impossible to take in, yet it prompted little action from a war-weary American people," wrote Karl Vick in a piece titled "ISIS

Leader's Rape of American Woman Sparks Little Outrage," shortly afterward in *Time*. "So if any shadow of a doubt remained about the limited American appetite for a new conflict, it was erased by the profoundly muted public response," of our leaders … and ourselves, I might add.

Comparisons between World War II and the current conflict aren't usually appropriate, as the people, situations, and times are so very different. But our current attitude and the volatile state of the world is increasingly reminiscent of how a war-weary West, broken and exhausted from the events of World War I, allowed the Nazis to rise to power, make great gains, and inflict unimaginable pain and suffering on the world before the righteous finally awoke from their pacifist dream.

Are we now asleep, dreaming of a peace that cannot exist—at least not in this age—without nations willing to fight against evil? Perhaps we are. If we ever wake up, I believe we'll still find that there are American warfighters willing to stand up to evil and send killers like those in ISIS to hell where they belong.

Meanwhile, a few thousand warfighters are now back in Iraq working as trainers, trying to teach the Iraqi security forces—again— how to fight, along with Syrians who'll confront ISIS in their homeland. Even though such a strategy has failed several times in the past, it's all the war-weary administration is willing to invest.

News made around the time of this writing indicates that the strategy isn't working. "We have to acknowledge that this is a total failure," said Senator Jeff Sessions during a September 2015 hearing examining our progress against ISIS. He was reacting to testimony from an American general who said that after $500 million and months of instruction, out of the fifty-four Syrians we trained, only "four or five" remained in the fight against ISIS; the rest were killed, captured, or ran away. Senator John McCain added that the Pentagon's assessment that the strategy is working was "divorced from reality."

Sadly, there seems no clear way ahead.

The warfighters profiled in this book are doing relatively well. I say *relatively* because most of them carry wounds, both physical and emotional, that we who didn't fight there cannot begin to understand. Most are them are still in the service and have progressed in rank, experience, and stature. Most returned to Iraq on second and even third deployments, or to Afghanistan. The moments chronicled in this book only recount a fraction of what they've done for this nation, and what they've done for each other. They haven't lived free from hardship, of course. Many of them think about the war every day, and several have seen their families suffer due to their long separations. Marriages have been strained, and divorce is common. They've all sacrificed much. Among the many things that we owe them, including the care and benefits we promised, is to follow through with what they fought so hard to achieve: victory. Allowing it to further slip away is a tragedy, and a great shame.

I interviewed each of the warfighters in this book several times between May of 2014 and June of 2015, during the time that ISIS rose and the U.S. returned to Iraq. When asked, none would comment on the political decisions made or predict the outcome of the current crisis, but all mentioned their disheartenment. "This is what we died for?" one of them said.

Justin LeHew is now a sergeant major and helps oversee one of the Marines' largest commands. As a testament to his reputation within the Corps, an obstacle is named in his honor on the course that all recruits must complete while undergoing basic training at Parris Island, South Carolina.

Eric Geressy became a sergeant major and retired after serving as the senior enlisted advisor at U.S. Southern Command. He keeps in close touch with his former soldiers, and remains employed in the defense community.

Jeremiah Olsen is a staff sergeant in the Army National Guard. He earned his bachelor's degree and is working on his master's with plans to become a physician's assistant. Ever the thrill-seeker, he talked with me one afternoon just after he returned from climbing up the side of a glacier.

Aubrey McDade is a gunnery sergeant and is back training young recruits who seek to become part of his beloved Corps. Every year on the anniversary of the Second Battle of Fallujah, he places several candles on a table in his home, along with his Navy Cross and a book about his battalion's experience in the war, to commemorate and remember the brave warfighters of B Company, 1/8 Marines.

Jarion Halbisen-Gibbs still wears the Green Beret. He's a master sergeant now, and was honored with a permanent display in Michigan's Military & Space Heroes Museum in Frankenmuth, Michigan.

Diem Tan Vo finally put that political science degree to use. He's a Foreign Service Officer with the State Department and also a major in the U.S. Army Reserve.

Chris Waiters is still a combat medic. A three-star general pinned the Distinguished Service Cross on his uniform and asked what assignment he would like. "I want to train new medics, sir," Waiters said. He had orders that afternoon to report to the combat medic school in Texas.

Joseph Miller medically retired from the Army and is studying to become a physical therapy assistant. "It's something that I can do that will allow me to continue helping people," he told me.

Jeremiah Church is currently undergoing the process to be medically retired, as well. After he endured years of incessant pain, doctors found bullet fragments in his lower back that have caused debilitating damage. "I was shot at so many times in Iraq and Afghanistan, I don't really know when they got there," he said. Church lives in the rural northeast and enjoys the outdoors with his wife.

Nick Eslinger became an Army Ranger, serving as a platoon leader in the same battalion that Jeremiah Olsen was in years earlier. The West Point Association of Graduates gave him a prestigious award for heroism in combat, and he plans to teach a course on leadership at the academy soon.

America will forever owe a debt of gratitude to these warfighters, to the thousands more they represent, and to their families who bore a burden for all of us. Our soldiers, sailors, airmen, Marines, and coastguardsmen represent the very best we have to offer. They come from all corners of our nation, from cities and farms, from manicured suburbs and down long country roads, from families who've been here since before our revolution to those who just arrived. Some come from privileged backgrounds; others were dirt poor. Most serve a few years and then seek the American dream, and a life of peace. A few become lifers.

All are members of a brotherhood where selflessness and courage are common virtues, and if ever our nation needs to fight and win a war again, we can rest assured that our American warfighters will be ready.

They've always been ready.

J. Pepper Bryars
March 2016
Alabama

BONUS MATERIAL

For exclusive bonus material, including unreleased chapters, additional photographs of the warfighters, and other excerpts, please visit:

www.jpepperbryars.com/americanwarfighterbonus

To receive offers and information on the latest writings from J. Pepper Bryars, please sign up for his newsletter at:

www.jpepperbryars.com/newsletter

A complete list of books written by J. Pepper Bryars can be found at:

www.jpepperbryars.com/books

SELECTED SOURCES

EPIGRAPH

Ambrose, Stephen. *D-Day: June 6, 1944: The Climactic Battle of World War II.* New York: Simon & Schuster, 1993.

Sledge, E. B. *With the Old Breed: At Peleliu and Okinawa.* New York: Presidio Press, 1981.

Simpson, Brooks D. and Berlin, Jean V. *Sherman's Civil War: Selected Correspondence of William T. Sherman, 1860–1865.* University of North Carolina Press, 1999.

CHAPTER ONE

"Support for War Spikes as Bush Sets a Deadline." *ABC News/Washington Post Poll,* March 18, 2003.

Al-Zubaidi, Ahmad. "Halabja: Survivors talk about horror of attack, continuing ordeal." *Radio Free Iraq,* March 15, 2008.

Bush, George W. "President's Remarks at the United Nations General Assembly." New York. September 12, 2002.

———. "President Bush Outlines Iraqi Threat." Cincinnati Museum Center – Cincinnati Union Terminal. Cincinnati, Ohio. October 7, 2002.

———. "Address to the Nation." The White House. March 17, 2003.

"Tough Talk from Toby Keith." *Country Weekly*, March 18, 2003.

Sides, Hampton. "The First to Die." *Men's Journal*, October 2003.

The New American Bible: Revised Edition. 2011.

The Iraq War: Regime Change. Dir. Charlie Smith. BBC. 2013.

CHAPTER TWO

Bayles, Fred et al. "Many of slain Marines saw Corps as life goal." *Associated Press*, March 25, 2003.

Blomquist, Brian. "Civilian hanged as Bush vows: We'll get fiends." *New York Post*, March 30, 2003.

Bush, George W. "Address to the Nation at the launch of the Operation Iraqi Freedom." The White House. March 19, 2003.

Docherty, Bonnie and Garlasco, Marc E. "Off Target: The Conduct of the War and Civilian Casualties in Iraq." *Human Rights Watch*, 2003.

James, Barry. "Marines' thrust meets little resistance: Blast at checkpoint kills 5." *New York Times*, April 5, 2003.

LeHew, Justin. Personal interviews. May 2014.

———. Narrative to Accompany the Award of the Navy Cross. U.S. Marine Corps. March 23–24, 2003.

Schmitt, Eric. "Army Chief Raises Estimate of G.I.'s Needed in Postwar Iraq." *New York Times*, March 30, 2003.

Slivka, Judd. "Valley Marine dies in Iraq battle." *Arizona Republic*, March 30, 2003.

CHAPTER THREE

"Saddam statue topples with regime." BBC, April 9, 2003.

Bush, George W. "President Bush Announces Major Combat Operations in Iraq Have Ended." USS Abraham Lincoln, at sea off the coast of San Diego, California. May 1, 2003.

Burns, John F. "A Nation at War; Tumult, Cheers, Tears and Looting in Capital's Streets." *New York Times*, April 10, 2003.

"Sahaf: U.S. troops will be burned." CNN, April 7, 2003.

DiManno, Rosie. "Hunt for the disappeared leads to hellish tunnels." *Toronto Star*, April 21, 2003.

Geressy, Eric. Personal interviews. October 2014.

———. Narrative to Accompany the Award for the Army Commendation Medal with Combat "V." U.S. Army, April 26, 2003.

Hari, Johann. "The Homecoming." *The Independent*, September 18, 2003.

Hendawi, Hamza. "U.S.: Forces Make New Raid Into Baghdad." *Associated Press*, April 6, 2003.

Hitchens, Christopher. *Hitch-22: A Memoir*. New York: Twelve, 2012.

Keegan, John. *The Iraq War*. New York: A. A. Knopf, 2004.

LeDuff, Charlie. "City Uncovers Book of Doom." *New York Times*, April 17, 2003.

"Liberated Baghdad." *Washington Post*, April 10, 2003.

Parsons, Tony. "Heroes who've inspired rebirth of national pride." *Daily Mirror*, April 7, 2003.

Ricks, Thomas E. "Army Historian Cites Lack of Postwar Plans." *Washington Post*, December 24, 2004.

CHAPTER FOUR

Bennett, Brian and Weisskopf, Michael. "The Sum of Two Evils." *Time,* May 25, 2003.

Bolger, Daniel P. *Why We Lost: A General's Inside Account of the Iraq and Afghanistan Wars.* New York: Houghton Mifflin Harcourt, 2014.

Bremer, L. Paul, and Malcolm McConnell. *My Year in Iraq: The Struggle to Build a Future of Hope.* New York: Simon & Schuster, 2006.

Chivers, C. J. "The Secret Casualties of Iraq's Abandoned Chemical Weapons." *New York Times,* October 14, 2014

Duelfer, Charles. "Comprehensive Report of the Special Advisor to the Director of National Intelligence on Iraq's Weapons of Mass Destruction." Central Intelligence Agency, September 30, 2004.

Gellman, Barton. "Frustrated, U.S. Arms Team to Leave Iraq." *Washington Post,* May 11, 2003.

Hitchens, Christopher. *Hitch-22: A Memoir.* New York: Twelve, 2012.

Kay, David. "Fox News Sunday with Chris Wallace." *Fox News,* February 1, 2004.

Olsen, Jeremiah C. Personal interviews. January 2014.

———. Narrative to Accompany the Award of the Silver Star. U.S. Army, June 11–12, 2003.

Sample, Doug. "The Faces Behind the Faces on the 'Most Wanted' Deck." *American Forces Press Service,* May 6, 2003.

Statement by the Press Secretary. "President Names Envoy to Iraq." The White House. May 6, 2003.

Tyson, Ann Scott. "Anatomy of the raid on Hussein's sons." *Christian Science Monitor,* July 24, 2003.

Wells, Chad. WWL-TV, Channel 4, New Orleans, August 2004.

CHAPTER FIVE

"24 killed in car bombing in Baghdad." *Associated Press,* January 19, 2004.

Camp, Richard D. *Battle for the City of the Dead in the Shadow of the Golden Dome, Najaf, August 2004.* Minneapolis, MN: Zenith, 2011.

"General: Despite bombing, U.S. has upper hand." CNN, February 10, 2004.

Fam, Mariam. "Car bomb at Iraqi army recruiting station kills dozens." *Associated Press,* February 12, 2004.

Graham, Patrick. "23 killed as Iraqi rebels overrun police station." *The Guardian,* February 14, 2004.

Harding, Luke and Haidar, Mohammad. "Iraq British-controlled Basra suffers its worst day since Saddam's fall as bombs destroy buses full of children." *The Guardian,* April 22, 2004.

LeHew, Justin. Personal interviews. May 2014.

———. Narrative to Accompany the Award of the Bronze Star Medal with Combat "V." U.S. Marine Corps. August 5–27, 2004.

MacAskill, Ewen. "Army of the dispossessed rallies to Mahdi." *The Guardian,* April 7, 2004.

Melson, Charles D. and Kozlowski, Francis X. "U.S. Marines in Battle: An-Najaf." U.S. Marine Corps History Division, August 2004.

Raphaeli, Nimrod. "Understanding Muqtada al-Sadr." *Middle East Quarterly,* Fall 2004.

Reid, Robert. "Iraq Bombing May Stiffen Kurdish Resolve." *Associated Press,* February 2, 2004.

CHAPTER SIX

Anderson, Nathan R. Obituary. *Akron Beacon Journal*, November 18, 2004.

"Body of blonde Caucasian woman found in Fallujah." *Agence France-Presse*, November 14, 2004.

Bremer, L. Paul, and Malcolm McConnell. *My Year in Iraq: The Struggle to Build a Future of Hope*. New York: Simon & Schuster, 2006.

Burns, John F. "Tape Condemns Sunni Muslim Clerics." *New York Times*, November 25, 2004.

Bush, George W. "Farewell Address to the Nation." The White House, January 15, 2009.

Epps, Matthew. *11-Mike: Memoirs of a Mechanized Infantryman*. 2013.

Filkins, Dexter. "In Fallujah, Young Marines Saw the Savagery of an Urban War." *New York Times*, November 21, 2004.

———. *The Forever War*. New York: Alfred A. Knopf, 2008.

McCarthy, Rory. "Uneasy truce in the city of ghosts." *The Guardian*, April 23, 2004.

McDade, Aubrey L. Personal interviews. June–July, 2014.

———. Narrative to Accompany the Award of the Navy Cross. U.S. Marine Corps. November 11, 2004.

McMahon, Colin. "U.S. promise to find killers, 'pacify' a city." *Chicago Tribune*, April 2, 2004.

Rhem, Kathleen. "Marine General: Fallujah Operations 'Ahead of Schedule.'" *American Forces Press Service*, November 11, 2004.

Ross, Brian. "Tracking Abu Musab al-Zarqawi." *ABC News*, September 24, 2004.

"U.S. declares insurgency 'broken.'" *Washington Times*, November 19, 2004.

West, Francis J. *No True Glory: A Frontline Account of the Battle for Fallujah.* New York: Bantam, 2005.

CHAPTER SEVEN

Beehner, Lionel. "Iraq: Insurgency Goals." Council on Foreign Relations, May 20, 2005.

Bush, George W. Remarks at the U.S. Army War College. Carlisle, Pennsylvania. May 24, 2004.

———. Remarks at the Commencement Ceremony at the United States Naval Academy. Annapolis, Maryland, May 27, 2005.

Cheney, Dick. *Larry King Live.* CNN, May 30, 2005.

Country Reports on Terrorism. U.S. Department of State, Office of the Coordinator for Counterterrorism. April 2005.

"English Translation of Ayman al-Zawahiri's letter to Abu Musab al-Zarqawi." *The Weekly Standard,* October 12, 2005.

Filkins, Dexter. "Defying Threats, Millions of Iraqis Flock to Polls." *New York Times,* January 21, 2005.

Getlen, Larry. "Could you be a green beret?" *New York Post,* January 6, 2013.

Glasser, Susan B. "'Martyrs' in Iraq Mostly Saudis." *Washington Post,* May 15, 2005.

Halbisen-Gibbs, Jarion. Personal interviews. March 2015.

———. Narrative to Accompany the Award of the Army Commendation Medal with "V" Device. United States Army. May 27, 2005.

Kyle, Chris, and Scott McEwen. *American Sniper: The Autobiography of the Most Lethal Sniper in U.S. Military History.* New York: W. Morrow, 2012.

"National Strategy for Victory in Iraq." National Security Council, November 2005.

Paz, Reuvan. "Arab volunteers killed in Iraq: an Analysis." Project for the Research of Islamist Movements, Global Research in International Affairs Center, March 4, 2005.

Roberts, Joel. "Poll: Fading Support for Iraq War." *CBS News*, October 10, 2005.

Tan, Michelle. "Spec ops needs 5,000 soldiers." *Army Times*, February 23, 2015.

CHAPTER EIGHT

Bolger, Daniel P. *Why We Lost: A General's Inside Account of the Iraq and Afghanistan Wars*. New York: Houghton Mifflin Harcourt, 2014.

Filkins, Dexter. *The Forever War*. New York: Alfred A. Knopf, 2008.

Geressy, Eric. Personal interviews. December 2014.

———. Department of Defense Form 338, Recommendation for Award: Soldiers Medal. U.S. Army, January 9, 2006.

Gordon, Michael R., and Bernard E. Trainor. *The Endgame: The Inside Story of the Struggle for Iraq, from George W. Bush to Barack Obama*. New York: Pantheon, 2012.

Losing Iraq. Dir. Michael Kirk. *Frontline*. PBS, July 29, 2014.

Ricks, Thomas E. "U.S. Counterinsurgency Academy Giving Officers a New Mind-Set: Course in Iraq Stresses the Cultural, Challenges the Conventional." *Washington Post*, February 21, 2006.

Worth, Robert F. "Blast at Shiite Shrine Sets Off Sectarian Fury in Iraq." *New York Times*, February 23, 2006.

CHAPTER NINE

Abizaid, John. Testimony. U.S. Senate Armed Services Committee. August 3, 2006.

Galula, David. *Counterinsurgency Warfare: The Theory and Practice.* New York: Praeger, 1965.

Hagel, Chuck. "Face the Nation with Bob Schieffer." CBS, August 6, 2006.

Kyle, Chris, and Scott McEwen. *American Sniper: The Autobiography of the Most Lethal Sniper in U.S. Military History.* New York: W. Morrow, 2012.

McMaster, H.R. *The Charlie Rose Show.* PBS, June 2, 2008.

Ricks, Thomas E. "The Lessons of Counterinsurgency." *The Washington Post,* February 16, 2006.

———. *The Gamble: General David Petraeus and the American Military Adventure in Iraq, 2006–2008.* New York: Penguin, 2009.

Smith, Neil and MacFarland, Sean. "Anbar Awakens: The Tipping Point." *Military Review,* March–April 2008.

Stack, Megan K. and Roug, Louise. "Fear of Big Battle Panics Iraq City." *Los Angeles Times,* June 11, 2006.

Vo, Diem Tan. Personal interviews. July 2014.

———. Narrative to Accompany the Award of the Silver Star. United States Army. September 27, 2006.

CHAPTER TEN

"33% Believe Troop Surge Will Work." *Rasmussen Reports,* January 29, 2007.

"57% Favor Immediate Troop Withdrawal or Firm Deadline." *Rasmussen Reports,* April 26, 2007.

ABC News / Washington Post Poll, January 10, 2007.

Bush, George W. "The New Way Forward in Iraq." The White House. January 10, 2007.

Macur, Juliet. "The Fobbits." *New York Times*, August 27, 2005.

Petraeus, David. Opening Statement. U.S. Senate Armed Services Committee Hearing. January 23, 2007.

Waiters, Christopher. Personal interviews. April 2014.

———. Narrative to Accompany the Award of the Distinguished Service Cross. U.S. Army. April 5, 2007.

Watanabe, Teresa. "A higher calling than duty." *Los Angeles Times*, February 16, 2007.

Zeleny, Jeff. "Leading Democrat in Senate Tells Reporters, 'This War Is Lost'" *New York Times*, April 20, 2007.

CHAPTER ELEVEN

"Poll: 19% Consider Troop Surge a Success." *Rasmussen Reports*, July 11, 2007.

Church, Jeremiah. Personal interviews. July 2013.

———. Narrative to Accompany the Award of the Silver Star. U.S. Army. August 8, 2007.

Cooper, Michael. "To Democrats, Report Proves War in Iraq Is Misguided." *New York Times*, July 19, 2007.

Kristol, William. "The Turn: Defeatists in retreat." *The Weekly Standard*, August 13, 2007.

Mansoor, Peter R. *Surge: My Journey with General David Petraeus and the Remaking of the Iraq War*. Yale University Press, 2013.

O'Hanlon, Michael E. and Pollack, Kenneth M. "A War We Just Might Win." *New York Times*, July 30, 2007.

Stolberg, Sheryl Gay and Zeleny, Jeff. "Clash Over Iraq Becomes Bitter Between Bush and Congress." *New York Times*, July 12, 2007.

CHAPTER TWELVE

Army Board of Correction of Military Records. Memorandum denying the appeal to upgrade the award of Silver Star to Distinguished Service Cross for Eric Geressy. April 16, 2015.

Blanks, Benjamin. Memorandum: Endorsement for upgrade from the Silver Star to the Distinguished Service Cross for SGM Eric J. Geressy. September 23, 2009.

Brook, Tom Vanden. "Pentagon may upgrade hundreds of troops to possible Medals of Honor." *USA Today*, January 6, 2016.

Butcher, Eric. "Iraq bombs: 250 die in worst terror attack." *The Telegraph*, August 16, 2007.

Geressy, Eric. Personal interviews. October 2014.

———. Narrative to Accompany the Award of the Silver Star. U.S. Army. September 4, 2007.

Huda, Metti. "Pontifical Babel College in Baghdad Finally Returned to the Chaldean Catholic Church." *Chaldean.org*, November 15, 2008.

Recommendation for Award of the Distinguished Service Cross for Sergeant Major Eric Geressy, DD Form 638. U.S. Army, August 3, 2010.

Turner, Christopher. Memorandum: Endorsement for upgrade from the Silver Star to the Distinguished Service Cross for SGM Eric J. Geressy. September 17, 2009.

"U.S. Ground Force End Strength." *GlobalSecurity*. August 2, 2015. Web.

Weber, James. Memorandum: Endorsement for update from Silver Star to the Distinguished Service Cross. September 21, 2009.

CHAPTER THIRTEEN

Bush, George W. "Remarks by the President at the 2008 Republican Governors Association Gala." National Building Museum, Washington, D.C. February 25, 2008.

Clinton, Hillary. Opening Statement. Joint Hearing of the Senate Armed Services and Foreign Relations Committees, September 11, 2007.

Halbisen-Gibbs, Jarion. Personal interviews. July 2014.

———. Narrative to Accompany the Award of the Distinguished Service Cross. September 10, 2014.

Kyle, Chris, and Scott McEwen. *American Sniper: The Autobiography of the Most Lethal Sniper in U.S. Military History*. New York: W. Morrow, 2012.

Mansoor, Peter R. *Surge: My Journey with General David Petraeus and the Remaking of the Iraq War*. Yale University Press, 2013.

"Measuring Stability and Security in Iraq. December 2007." Department of Defense Report to Congress. December 2007.

Moveon.org. "General Petraeus or General Betray Us?" Advertisement. *New York Times*, September 10, 2011.

Obama, Barack. Opening Statement. Joint Hearing of the Senate Armed Services and Foreign Relations Committees, September 11, 2007.

Petraeus, David. "Report to Congress on the Situation in Iraq." Joint Hearing of the House Armed Services and Foreign Affairs Committees, September 10, 2007.

"Poll: U.S. split over Afghan troop buildup." CNN, November 24, 2009.

Shanker, Thom. "The Time Has Come, the General's Here: Petraeus Preps for Testimony on Iraq." *New York Times*, September 8, 2007.

CHAPTER FOURTEEN

Eslinger, Nick. Personal interviews. October 2014.

———. Narrative to Accompany the Award of the Silver Star. U.S. Army. October 1, 2008.

Filkins, Dexter. "U.S. Hands Off Pacified Anbar, Once Heart of Iraq Insurgency." *New York Times*, September 2, 2008.

"2 Most Wanted al-Qaeda Leaders in Iraq Killed by U.S., Iraqi Forces." *Fox News*, April 19, 2010.

Gates, Robert. "Remarks at Multi-National Force-Iraq Change of Command Ceremony." Iraq. September 16, 2008.

Muradian, Vago. "Gen. James Mattis, U.S. Joint Forces Command." *Defense News*, May 23, 2010.

Obama, Barack. "Remarks Against the Iraq War." Chicago. October 2, 2002.

———. "Responsibly Ending the War in Iraq." Camp Lejeune, North Carolina. February 27, 2009.

Petraeus, David. "Letter to the Soldiers, Sailors, Airmen, Marines, Coast Guardsmen, and Civilians of Multi-National Force-Iraq." September 15, 2008.

Shadid, Anthony. "Iraqi Insurgent Group Names New Leaders." *New York Times*, May 16, 2010.

Rubin, Alissa. "Iraq Marks Withdrawal of U.S. Troops From Cities." *New York Times*, June 30, 2009.

CHAPTER FIFTEEN

Arando, Tim and Schmidt, Michael E. "Last Convoy of American Troops Leaves Iraq." *New York Times*, December 18, 2011.

Cooper, Helene and Stolberg, Sheryl Gay. "Obama Declares an End to Combat Mission in Iraq." *New York Times*, August 31, 2010.

Fischer, Hannah. "A Guide to U.S. Military Casualty Statistics: Operation Freedom's Sentinel, Operation Inherent Resolve, Operation New Dawn, Operation Iraqi Freedom, and Operation Enduring Freedom." Congressional Research Service. August 7, 2015.

Garamone, Jim. "Obama Praises U.S. Troops as Iraq Winds Down." *American Forces Press Service*, December 14, 2011.

Geressy, Eric. Personal interviews. August 2015.

Halbisen-Gibbs, Jarion. Personal interviews. August 2015.

Jakes, Lara and Santana, Rebecca. "Iraq Withdrawal: U.S. Abandoning Plans to Keep Troops in Country." *Associated Press*, October 15, 2011.

Londono, Ernesto. "Operation Iraqi Freedom ends as last combat soldiers leave Baghdad." *Washington Post*, August 19, 2010.

"Measuring Stability and Security in Iraq. August 2010." Department of Defense Report to Congress. August 2010.

Obama, Barack. "Remarks by the President and First Lady on the End of the War in Iraq." Fort Bragg, North Carolina. December 14, 2011.

Panetta, Leon. "How the White House Misplayed Iraqi Troop Talks." *Time*, October 1, 2014.

Zucchino, David. "In Iraq, U.S. troops aren't yet in the clear." *Los Angeles Times*, December 1, 2011.

EPILOGUE

Aridi, Sara. "Why Kayla Mueller, raped repeatedly by ISIS leader, refused to escape." *Associated Press*, August 15, 2015.

Baldor, Lolita and Riechmann, Deb. "General: Only handful of Syrian fighters remain in battle after many killed or left the fight." *Associated Press*, September 16, 2015.

Callimachi, Rukmini. "ISIS Enshrines a Theology of Rape." *New York Times*, August 13, 2015.

"General Admits We've Only Trained 'Four or Five' Anti-ISIS Fighters." *Daily Caller News Foundation*, September 16, 2005.

Griffin, Jennifer and Tomlinson, Lucas. "Army chief Odierno, in exit interview, says US could have 'prevented' ISIS rise." *Fox News*, July 22, 2015.

Holdaway, Owen and Hall, John. "U.S. hostage Kayla Mueller had her fingernails pulled out before being repeatedly raped by ISIS leader: Yazidi sex slave reveals torment of aid worker's harrowing final months as the secret wife of al-Baghdadi." *Daily Mail*, September 9, 2015.

McCarter, Mark. "'Perverted jihadist' won't shake military's resolve, Biden says." *Huntsville Times*, August 16, 2015.

Petraeus, David. "Remarks at the Aspen Ideas Festival." Aspen, Colorado. June 30, 2014.

Vick, Karl. "ISIS Leader's Rape of American Woman Sparks Little Outrage." *Time*, August 24, 2015.

ACKNOWLEDGEMENTS

Vaughn, Sharon. "My Heroes Have Always Been Cowboys." Perf. Nelson, Willie. Columbia, 1980.

ACKNOWLEDGMENTS

I would like to thank my wife, Rachel, for her encouragement, advice, edits, and unwavering belief in my vision for this book. It wouldn't exist without her.

I would also like to thank my five children, who motivate me to become a better man, and that includes a better provider and writer.

I give my thanks to the Lord, who I believe guided me to these men so that their stories could be told.

There are many individuals who have contributed to this book, in ways large and small, recently and a long time ago: my mother, Dianne Smith Bryars, who once gave me a biography of the great war correspondent Ernie Pyle, and who always encouraged me to write a book like this. My journalism professor at Spring Hill College, the late Pat McGraw, who introduced me to the work of Studs Terkel. His oral-history writing style inspired this book. The noncommissioned officers of the 131st Mobile Public Affairs Detachment, Alabama Army National Guard—Mike McCord, Chris Brown, Jamie Brown, and Taylor Barbaree—who all taught me how to write about the military. Mike Marshall, a wise editor who gave me my first job as a writer.

Governor Bob Riley, who gave me an opportunity to see how Congress and a governor's office worked from the inside. Gary Thatcher and Dan Senor, who brought me aboard the Coalition Provisional Authority's staff in Baghdad, where I saw the war firsthand and learned more than I ever could from books or articles. Ray DuBios, who offered me an appointment as a Defense Fellow in the Office of the Secretary of Defense during a critical period of the war. And Noreen Holthaus, whose thoughtful leadership allowed me to learn a great deal about the war while serving at the Pentagon.

There are many others, of course, and I thank you all.

Finally, my eternal appreciation goes to the warfighters in this book, and their families. Thank you for entrusting your stories to me. To allude to an old country song, my heroes have always been soldiers ... and they still are, it seems.

ABOUT THE AUTHOR

J. PEPPER BRYARS began his career writing for military newspapers while serving in the Army National Guard, and he received the Army Commendation Medal for his deployment to Hungary in support of the peace-enforcement mission in Bosnia.

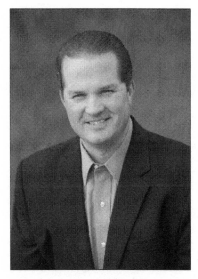

Pepper then became a newspaper reporter, spent time as an aide to a congressman and governor, and served as a presidential appointee in the Defense Department. He was also a strategic communication advisor to U.S. military forces operating in Europe, Africa, and Latin America. He was twice awarded the Office of the Secretary of Defense Award for Exceptional Public Service, once for service in Baghdad and a second time for work at the Pentagon.

His weekly opinion column is published in the *Birmingham News*, *Mobile Press-Register*, *Huntsville Times*, the *Mississippi Press*, and at AL.com.

Readers may contact him at www.jpepperbryars.com.

Made in the USA
San Bernardino, CA
05 September 2017